Managing Breeds for a Secure Future: Strategies for Breeders and Breed Associations

D. Phillip Sponenberg, DVM, PhD
Donald E. Bixby, DVM

The American Livestock Breeds Conservancy
Pittsboro, North Carolina

The American Livestock Breeds Conservancy
PO Box 477, Pittsboro, North Carolina 27312 USA

Managing Breeds for a Secure Future:
Strategies for Breeders and Breed Associations

© 2007 by the American Livestock Breeds Conservancy
All rights reserved. Published 2007
First printing 2007

No part of this publication may be reproduced, stored in a retrieval system, or transmitted in any form or by any means, except for short excerpts used in reviews, without the written permission of the authors or copyright owner. Requests for permission to reproduce parts of this work, or for more information, should be addressed to the American Livestock Breeds Conservancy.

The information presented in this book is true and complete to the best of our knowledge. The authors and publisher disclaim any liability in connection with the use of this information.

D. Phillip Sponenberg 1953 –
Managing Breeds for a Secure Future: Strategies for Breeders and Breed Associations / by D. Phillip Sponenberg and Donald E. Bixby
Includes index: Under investigation

ISBN 1-887316-07-8

Library of Congress Catalog Card Number: 2007938298

Cover: Torsten D. Sponenberg

Acknowledgements

Fred Horak deserves great credit for pushing this compilation of ideas and approaches into being. Without his gentle prodding this book would have never seen the light of day. It is our goal that breeders of several breeds will benefit from his persistence. The staff of the American Livestock Breeds Conservancy have been very, very helpful over the years in honing our thinking concerning breeds and their importance to the agricultural landscape. Without Carolyn Christman, Marjorie Bender, Libby Henson, Laurie Heise, Don Schrider, Chuck Bassett, Cindy Rubel, Anneke Jakes, and Michele Brane this work would never have been possible. Don Schrider was very helpful in finding the last few photographs that help to illustrate this book. Marjorie Bender is especially appreciated for her good humor as well as her phenomenal ability to point out both strengths and weaknesses in any written presentation. The work of ALBC has been pivotal in reversing the trend of breed erosion in the USA, and a great debt of gratitude is owed to all ALBC members and friends involved in this important work. While several have helped to spur this book into being, all weaknesses and errors are ours alone.

Torsten Sponenberg gave the book a helpful final edit, and her suggestions have made for easier reading and understanding. Her ability to counteract confusing wording as well as to suggest simpler ways to present detailed concepts have greatly helped the book.

* * * * *

Funding for this publication was generously provided by Cedar Tree Foundation as part of a grant for the Renewing America's Food Traditions (RAFT) Project to the American Livestock Breeds Conservancy, Chefs Collaborative, Center for Sustainable Environment, Culture Conservancy, Native Seeds/SEARCH, Seed Savers Exchange, and Slow Food USA.

The American Livestock Breeds Conservancy

FOREWORD

The goal of this book is to lay out successful strategies for breed survival and the important role of breed associations as breed advocates, and it results from our pondering and discussing many thorny issues that have arisen over decades of working with livestock breed conservation. Nearly all of this work has been undertaken with the American Livestock Breeds Conservancy (ALBC), founded in 1977. ALBC has emerged as a globally respected source for information and procedures concerning rare breed conservation. Individual breeds as well as individual breeders have often alerted ALBC to situations for which clear-headed responses were needed to assure survival of important genetic resources.

Responding to those challenges and needs over a period of many years has generated strategies that have gradually been distilled to a general approach that succeeds in most situations to assure breed survival and utility, while also being practical for breeders. Each breed of livestock encounters unique challenges and needs, though they all have a great deal in common so that certain broad strategies will succeed for many of them.

The goal of this book is to lay out successful strategies for breed survival.

This book delves into the "why" of breed conservation, as well as the "how" aspects. Our goal is to stimulate thought on the important issues of breed organization for conservation success so that readers can base specific decisions on a sound guiding framework of philosophy and goals.

Contents

Acknowledgements iii
Foreword v
1. Breeds 1
    Sustaining Breeds Over Time 5
2. Biology of Breeds 7
    Classes of Breeds 9
        Landraces 9
        Standardized Breeds 14
        Modern "Type" Breeds 16
        Industrial Strains 17
        Feral Populations 18
        A Review of Classes of Breeds 19
        What Should Be Included in a Breed 20
        One Breed or Two? 23
    Breeds as Gene Pools: Variability and Predictability 25
    Genetic Organization of Breeds 27
        Bloodlines within a Breed 34
    Breed Histories 35
    Geography and Source Herds 37
3. Breed Standard.. 39
    The Development of a Breed Standard........................... 42
    Breed Standards and Genetic Diversity 44
    Breed Type 47
    Qualitative and Quantitative Traits 50
    Changes to the Breed Standard 51
    Breed Type Reproduces Breed Type 52

4. Maintaining Breeds 53
Breeding Strategies 53
Linebreeding and Inbreeding 53
Outcrossing: Crossbreeding and Linecrossing 56
Defining Matings as "Related" or "Unrelated" 59
Linebreeding or Outcrossing: Which Is Best? 60
Linebreeding and Linecrossing as
    Strategies for Population Management 65
Important Applications of Linebreeding 67
Inbreeding and Loss of Diversity 69
Monitoring Inbreeding 71
Monitoring Effective Population Size 73
Generation Interval 75
Inbreeding within Individual Herds 77
Inbreeding within Breeds 77
Combining Linebreeding and Linecrossing 79
Inbreeding and Outbreeding Summary 80
Over-representation of Individual Animals 80
Genetic Bottlenecks 83
Artificial Insemination, Embryo Transfer,
    Assisted Reproductive Technologies 84
Mating Systems and Pairing of Animals within Purebred Breeds 84
Selection 87
Degree of Selection 88
Selection and Type 91
Selection of Animals for Reproduction 92
Use of Estimated Breeding Values 94
Gene Flow Into and Out of Breeds 96
Upgrading and What It Does 97
Upgrading and Bloodlines 105
Upgrading to Rescue Endangered Bloodlines 106
Adaptation 107
Long-term Genetic Management of Breeds 110

5. External Factors Affecting Breeds 111
Market Demand 111
Crossbreeding 115
Regulations 116
Imports 118

    Imports That Contribute Substantially to Conservation Efforts 118
    Imports That Enhance American Bloodlines 120
    Imports That Hamper Conservation in the Country of Origin 121
    Imports That Endanger American Bloodlines and Breeds 122
    Assessment of Importations 126

6. Associations 127
    Purposes of Associations 128
    Communicating the Association Purposes 128
    Membership 129
    Breed Associations for Endangered Breeds 130
    Communication 131
    Communication Networks within Associations 131
    Multiple Breed Associations 133
    Codes of Ethics 135
    Educational Programs 136
    Research 138
    Recruiting and Training New Breeders 139
    Breed Promotion 142
        Breed Sale Events 143
    Forms of Association 144
        Private Associations 144
        Unincorporated Associations 146
        Incorporated Associations 146
        Bylaws 147
        Board of Directors 148
        Directors and Officers 149
    Networks of Breed Associations 150
        Species Associations 150
    Promoting the Association 151
    Association Reputation 151
    Association Responsibilities 152
        Conservation Responsibilities 152
        Providing Pedigree Information 152
        Providing Breed Health Status Reports 152
        Reporting Measures of Genetic Diversity 155
        Development of Programs to Save Herds in Peril 155
        Development of Long Range Conservation Plans 156
        Dispelling False Rumors Quickly 157

  Conflict of Interest 157
  Local and Regional Groups 158
7. Competitive Shows 159
  Card Grading 161
  Non-competitive Exhibition 163
8. Registry 165
  Registration 165
  Pedigrees 166
  Obtaining Pedigree Information 168
  Pedigrees and Genetic Diversity 168
  Pedigree Recording Systems 169
  Litter Recording 170
  Stud Reports 171
  Selective Recording Systems 171
  Registrations Are Important 172
  Closed Herd Book Registries 174
  Open Herd Book Registries 175
  Registration of Crossbreds and Partbreds 178
9. Monitoring Breed Populations 179
  Monitoring Breed Health 182
  Congenital and Genetic Defects 182
10. Breeder Responsibilities 185
  Breeds, Breeders, Associations, and the Future 186
Appendices 189
  Appendix 1. Colonial Spanish Horse Score Sheet 189
  Appendix II. Sample Bylaws 192

1. Breeds

Breeds are important subdivisions that exist within domesticated species. Breeds are somewhat similar to the subspecies classification that is applied to species of wild animals. Both breeds and subspecies are groups of animals with significant and distinctive similarities that set them apart from other animals of the same species. While the word "breed" is frequently bandied about, it is in fact very difficult to define in a meaningful and consistent way. This difficulty arises because of two major factors that affect breeds and their function. One factor is the biological aspect and results from the role of a breed within its species. The other factor is political and arises from the important interactions of breeds as they function in human culture and under human management. As the biological and political aspects meet and interact they can take the definition of "breed" to different extremes. The least restrictive definition is that a breed is a "group of animals with a fence around it." This implies that breeds have some degree of genetic isolation, as well as some degree of identity. Another powerful definition of "breed" is a group of animals whose individual members resemble each other closely enough to be readily recognized, and that reproduce this same breed type when mated together. This definition implies an agreement among a group of breeders about the characteristics that define the breed. It is this definition that is adopted and employed throughout this book.

A breed is a group of animals that are similar and reproduce their same type when mated together.

Livestock breeds are fascinating biological entities with a profoundly important, and growing, role in maintaining global biodiversity. The practical significance of breeds springs largely from their unique roles in agricultural ecosystems. Breeds exist only in domesticated species, and the agricultural setting is the basis for the importance of breeds in most species. Domesticated species occupy unique and peculiar biological spaces where both natural and human influences meet. As human populations have exploded over the centuries, agricultural environments have increasingly expanded and diversified and are now large and pivotal components of the global ecosystem. Assuring that

2   Breeds

these environments function sustainably is critically important to the future of the earth and its human and nonhuman inhabitants.

Nearly all domesticated species are subdivided into many subpopulations, which in most situations are called breeds. Each breed functions in a different agricultural and cultural environment, and the roles that the different breeds play in these environments can have profound environmental effects. The interaction of any breed with its environment also shapes its genetic character, which is essential to its function. For most livestock species, breeds serve as the main reservoirs of the biodiversity within the species. Half of the biodiversity of most domesticated species is shared across breeds, while the other half is unshared and is instead contained only within single breeds. The consequence of this structure of genetic diversity is that losing breeds means losing biodiversity, because by losing a breed the species loses the genetic information that is unique to that breed.

Half of the genetic diversity in most species is shared across breeds; the other half is unshared and only within a single breed.

For example, the Leicester Longwool sheep breed is adapted to wet lowland conditions, while the Merino breed is adapted to dry grasslands. These two breeds vary from one another in a host of genetic differences that separate them by wool type, disease resistance, forage use, and behavior. Many of the variants of one of these breeds are simply lacking in the other, so to lose one choice or the other through breed extinction would drastically reduce choices for future sheep breeders.

The importance of breeds within domesticated species can hardly be overstated. This is especially true for those species, such as cattle and horses, whose wild progenitors are now extinct. Cattle and horses are now uniquely agricultural rather than wild, and consequently the various breeds embody the entire

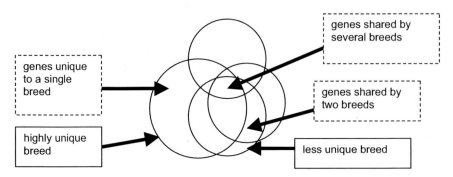

The genetic diversity in livestock species is contained in breeds.

genetic diversity of these species. To lose this diversity is to assure that unique components of global biodiversity are lost forever because it is now impossible to turn to the wild ancestors of these species for their genetic material; they are extinct. Effective management of breeds is essential to managing the biodiversity of all domesticated species, and is an issue that will only gain importance as globalization increasingly shrinks the world and its choices.

Losing breeds means losing biodiversity.

Agricultural environments have been vital sustainers of human life and cultural development for over 10,000 years. As agriculture has expanded around the globe it has had an increasingly large effect on the overall global ecosystem. Agriculture is the base from which humans dramatically increased in number and became the major determinant of global ecology. Throughout most of human history agriculture has been viewed as a necessary contributor to human sustenance and has been highly valued in that role. Agriculture has had great and lasting success in those regions where it was viewed as a sustainable and renewable system to be carefully stewarded. In contrast, mainstream modern thinking, and indeed some thinking long past, has viewed agriculture as an extractive and exploitative endeavor, similar to mining. This mode of thought quickly leads to non-sustainable systems, which cannot continue to remain productive over multiple centuries without massive inputs from outside the agricultural system itself. Fortunately, agricultural thinking has begun to return to a more renewable and sustainable mindset. It is within that mode of thinking that breed diversity best functions and flourishes. Understanding the relationship of breeds to varied agricultural settings is key to appreciating and fostering their

The extinct wild Auroch persists today only as domesticated cattle.

performance and survival.

Breeds of livestock are important components of any renewable and sustainable agricultural system. Breeds are diverse and each functions best in a specific environment and under a specific management strategy. The fit of each breed into a specific environment and for specific production goals is vital to appreciating the value of the diversity of breeds. Breeds that are removed from their production environments and are kept only as interesting relics by dedicated hobbyists lose much of their powerful contribution to human endeavor and lose much of their relevance. They indeed also lose much of their genetic makeup due to the change in selection pressure. Genetic characteristics are likewise lost when all breeds within a species are selected to function equally across all environments for a single production goal, for they then lose much of the genetic distinctiveness that originally contributed to their agricultural relevance in diverse settings.

Breed survival fails when biological or political influences are mismanaged.

Breeds, as a consequence of their development in agricultural environments, are the product of both natural- and human-generated influences, and both are essential for their survival. Breeds fail to survive when either the biological or the political influences are ignored or are mismanaged. This book brings together the biological aspects of breed function along with the politics and human organization of breeds. Success in both arenas is necessary to assure the persistence of breeds as critically important components of global biodiversity. Understanding why to conserve breeds and how to conserve breeds are both important for conservation to succeed. Both "why" and "how" will be detailed in the following chapters.

The Pineywoods environment shaped the resistant and adaptable Pineywoods cattle breed. Photo by D. P. Sponenberg.

Improved grasslands are productive and provide an environment that shapes productive breeds such as this Red Poll cow. Photo by Nathan B. Melson.

A specific note on nomenclature is warranted at this early point. Populations of various species of animals are known by different terms: *herds* of cattle, horses, goats, and swine, *flocks* of sheep, ducks, chickens, and turkeys, and *gaggles* of geese. Throughout this work the usual word chosen is "herd," although this is meant to be inclusive of all species while at the same time avoiding the distracting repetition of "herd/flock/gaggle" when describing groups of animals. Breeders interested in species that do not occur in herds are urged to look past the limitations of nomenclature to see the overarching truths that impact the management of breeds of all species.

Sustaining Breeds Over Time

Breeds can be sustained as viable biological entities over long periods of time, although this takes forethought and wise action. Breeds can only be sustained if their biological character and political aspects are both managed for success. In nearly every case a major factor in breed prosperity is steady demand for the breed, usually because of steady demand for its products. In a situation of steady demand, breed rarity is avoided.

Breed-specific production systems can be important for breed stability and sustainability. Changes in a production system can dictate changes in selection pressures, and these cannot avoid changing the genetic makeup of any breed. In the process of change, some genetic information is irretrievably lost. This loss, it is hoped, can be managed to occur in such a way that little is irretrievably lost.

In addition to the genetic repercussions of production system changes are the more subtle sociological changes that accompany them. When systems change, so do the cultural trappings that are associated with animal use and maintenance. Much of the vocabulary, procedures, and fine details of animal management are part of a rich culture that is only rarely conveyed in writing. As a result, failure of continuity for even one generation assures that certain cultural practices are lost. A common pattern is that families leave farming in search of better economic opportunities. Later generations of the same family may then return to agriculture, if only as a hobby or part-time endeavor. While much is saved by this approach, many aspects of the culture and practice of animal production are lost because the intervening non-agricultural generation has failed to provide the continuity that such cultural exchange needs. Even simple details like how to call the cows home can be lost forever if only a single generation fails to learn how to do it from the long line of generations that have called them home in the same way for centuries. Such lapses impoverish the richness of the culture of animal production systems.

The threats to the continuity of the cultural aspects of animal production have no easy answer. It is much easier to manage and save the genetics of breeds than it is the traditional systems under which they have evolved. It is worth pondering, though, that saving traditional systems will nearly always result in saving the genetic resources (breeds) that they spawned. In contrast, it is much more difficult to effectively save the intact genetic resources in the absence of the systems responsible for their origin and persistence.

The culture of animal use and production is best accomplished by assuring continuity of the systems surrounding them. Details of ox training and use, for example, are quickly lost when a single generation stops using oxen such as these Pineywood oxen. Photo by D. P. Sponenberg.

2. BIOLOGY OF BREEDS

One of the most useful definitions of a "breed" is one that is strictly biologically based. Juliet Clutton-Brock first articulated this definition, which asserts that a breed is a group of animals that is consistent enough in type to be logically grouped together, and that when mated within the group reproduces the same type (Clutton-Brock, Juliet. 1987. *A Natural History of Domesticated Mammals*. London, Cambridge University Press and British Museum of Natural History). This definition very ably gets to the core of the biological aspect of the breed concept: breeds are consistent and predictable genetic entities. The status of breeds as genetic resources is a consequence of the fact that breeds breed true.

One internationally accepted definition of "breed" is that a breed is any animal population deemed to be a breed by the governmental authorities in its region of occurrence. This is a very loose and inclusive definition that is derived more from political expediency than from biological principles, and it governs animal breeding in most countries. By this definition, a breed is anything that breeders say it is, and as a result some variable populations, such as Pinto horses, are included as breeds along with other more genetically defined breeds. The biologically based definition is much more strict, but is also more useful because it targets identification (and, it is hoped, conservation) to those populations most likely to make significant contributions to biodiversity.

Breeds are predictable genetic entities: breeds breed true.

Breeds serve human needs in agricultural systems because they are readily identifiable genetic packages, and each package is repeatable among members of the breed. This repeatability serves to make breeds predictable, and this is their main importance. The individuals of a breed are much more than an assembly of individual genetic variants of different utility to humanity, but are instead a complete package of specific genetic variants. The entire genetic package of specific combinations of genetic variants is much more fundamental to the significance of breeds than are the individual genetic variants, however unique or interesting some of those may be. The philosophy of breeds as packages of genetic combinations rather than as repositories for single genes leads to the holistic conclusion that breeds, rather than just their individual component genes, need to be conserved for their potential use in agricultural systems. This philosophy is useful in guiding conservation and management strategies.

The biologic definition of breed best serves conservation purposes. Many collections of animals are designated as "breeds" but fall far short of genetic breed status if the biological definition is used. These are usually groups of animals that exhibit a superficial feature in common such as spotting on Pinto horses that vary in type from miniature to draft, or an external phenotypic similarity that masks great genetic diversity such as in some Warmblood horse breeds. These loose definitions of breeds do indeed serve as useful designations for breeders interested in animal populations with certain characteristics, but the genetic variability of these populations reduces their predictability and precludes their usefulness in a conservation sense.

Only genetically based breeds are predictable genetic resources.

Most animal populations fall somewhere along a continuum from very uniform and predictable genetic packages to completely variable and unpredictable genetic packages. Exactly where to draw the line between the genetic "breed" and "nonbreed" is somewhat arbitrary and frequently controversial. The concept that some culturally-defined breeds fall outside of the realm of genetically-defined breeds is important. Ignoring this fact can result in conservation efforts being squandered on populations that have little to offer as genetic resources.

Breeds vary immensely, and each is a unique package of genetic variants. While Milking Shorthorns and Dexters are both dual-purpose breeds, their differences allow them to be successfully matched to very different production systems. Photo by D. E. Bixby.

Only breeds that qualify as breeds by the genetic criterion are useful as predictable genetic resources, and it is these that should be especially targeted for wise long-term management and conservation.

Classes of Breeds

Within the large group of breeds that satisfy the biological definition of breed (and some that do not) are a few major classes of populations. Each class has a different genetic history, and these differences have important consequences for effective maintenance and conservation. The major classes of breeds are landraces, standardized breeds, modern "type" breeds, industrial strains, and feral populations. Each of these classifications reflects differences about the attitude of human caretakers towards breeds as genetic packages, and each has something to teach about breeds, genetic packages, human endeavor, and the interaction of human management decisions and the genetic structure of animal populations.

Classes of breeds are:
- *Landraces*
- *Standardized Breeds*
- *Modern "Type" Breeds*
- *Industrial Strains*
- *Feral Populations*

Landraces

Landrace breeds represent an early stage of breed development. "Landrace," as used here, is a general term that refers to populations of animals that are isolated to a local area where local production goals and the physical environment drive selection. The "landrace" designation should not be confused with the specific Landrace swine breed, nor with the Finnish Landrace sheep breed (now known as Finnsheep). The landrace concept, as used here, is important as a general pattern for many breeds of all species. Landraces are sometimes called local breeds or natural breeds.

Landraces derive their unique genetic character from a combination of founder effect, isolation, and environmental pressures. The usual history of landraces begins with introduction of animals from some specific source (generally whatever was at hand) into a specific geographic region with its unique climate, forage, and topography. In most cases this was followed by isolation from further introduction of genetic material. Most landraces have then faced survival pressures of a compromised environment outside of the agricultural mainstream. Founding event, subsequent isolation, and the selection environment are the three basic factors that combine to determine the overall type and function of landraces. All three are important, and all three must be considered when developing effective conservation programs for these unique and useful genetic resources.

Landraces achieve their genetic consistency and uniqueness by default rather than by design, even though human selection for production goals is typical of most landraces. Founders of landraces usually emerge as the result of an accident of history rather than the result of a careful and deliberate selection process. A useful example of this phenomenon is the family of Criollo cattle breeds of the Americas.

Criollo cattle populations were founded by Spanish explorers from cattle brought to the New World from southern Spain. These cattle were selected by virtue of convenience – they were the closest to the ports of departure in the late 1400s and early 1500s. Spaniards took fewer than 300 head of these cattle to the Americas. These few cattle provided the genetic basis for breeds as geographically far-flung as the Texas Longhorn, Florida Cracker, Pineywoods, Romo Sinuano of Colombia, Criollo of Argentina, and a multitude of breeds in between. All of these breeds are similar by virtue of their close relationship to the original few cattle (founder effect). Their differences come from centuries of divergent selection, isolated from one another in diverse climates and topographies. The founder package, however, forever constrains the overall range of types that is possible in this important family of breeds, and it is still possible to find many individual

Founders, isolation, and environment combine to create unique landrace breeds.

This Pineywoods cow could easily be mistaken for a Criollo cow from any of several breeds throughout the Americas, because founder effect has made all of these breeds similar to one another. Photo by D. P. Sponenberg.

cattle in nearly all Criollo breeds that are easily mistaken for other breeds within the group.

Following the founding of animal populations, isolation is important for the development of landraces as genetic resources. In contrast, lack of isolation allows for repeated introduction of new genetic variants resulting in a population that never achieves the genetic consistency expected of a breed. Isolation for most landraces has been due to geographic factors. Relative lack of communication and transportation infrastructure has also been important in the establishment of genetic isolation for many landraces. As history progresses and global development proceeds, the isolation that long protected these genetic packages is disappearing. With decreasing isolation the uniqueness of many landraces is diminishing, and with their disappearance go many highly adapted and potentially useful genetic characteristics. Isolation is critically essential for landraces, as they will otherwise fail to be genetically uniform to the extent that is necessary for them to serve as useful genetic resources.

Isolation is critically essential for landraces.

Early isolation of landraces was much more by default than by design. It is increasingly risky to rely on default to isolate landraces because improvements in communication and transportation remove the barriers to the importation of exotic breeds into a landrace's home range. The temptation to crossbreed landraces out of existence is generally more than can be resisted by the caretakers of these genetic resources. Crossing can provide a quick, if temporary, financial return, and can frequently become the strategy for the best short-term outcome. Unfortunately, the long-term outcome of such crossbreeding is usually below the productive potential of the original uncrossed landrace.

Most landraces lack tightly organized breeder organizations. Lack of organization works well when isolation is assured by limited infrastructure. As soon as infrastructure improves, though, any nondeliberate isolation fails to function very well and more formal strategies for isolation and organization of landraces must be developed if these genetic resources are to survive.

The very cultural setting that once provided for the development and maintenance of landraces is now itself largely gone and can no longer foster the animal genetic resources that it spawned. Conservation must now be intentional rather than by default. The biggest hurdle to overcome with landrace conservation is the fact that the cultural and physical space in which they developed has changed dramatically over the last several decades. Isolation that protected them has diminished so that it no longer assures protection. As long as cultural, geographic, and infrastructural barriers isolated landraces, they were safe in their original habitat and for their original purposes. With the relative homogenization of culture both nationally and internationally in the last few decades, the unique and peripheral cultural pockets in which landraces persisted are now largely gone. In addition, landrace populations have been resettled in new cultural and environmental habitats where historic selection pressures are different than in their original ranges. To rely on this modified agricultural environment to save landraces is to assure their extinction. Conservation must therefore result from carefully crafted intentional efforts rather than by the default of isolation that can no longer be realistically achieved.

Examples of radically changed environments include those for both Pineywoods and Texas Longhorn cattle. Both breeds were originally landraces that thrived in vast open-range systems. As society and expectations changed, land became subdivided by fences, and more deliberate mating choices were made. Breed type and breed expectations changed along with irreversible changes in the management and physical environment for both of these breeds.

The Texas Longhorn went from a wide-ranging adapted animal in extensive systems to one that was more likely to be cared for in a small, carefully tended herd. The selection pressures became much more focused on looks than on survival, and as a result cattle in the breed became larger, smoother, and more dramatically colored than was typical of the 1800s when the breed had its widest distribution. Important exceptions to the general rule exist, but the general character of the Texas Longhorn breed has changed much since the 1800s.

Gulf Coast sheep are a landrace that is imperiled by improved communication and transportation. Photo by D. P. Sponenberg.

Breeders of Pineywoods cattle face the challenge of balancing adaptation with improved production characteristics. Photo by D. P. Sponenberg.

Pineywoods cattle are at a much earlier stage of modification than are Texas Longhorns, due in part to the persistence of the open range for much longer in the Deep South than in Texas. The Pineywoods cattle breeders are much more aware of the potential threat of breed uniformity and the loss of variation in the breed, and are likely to succeed in conserving this breed in a form that has persisted for centuries.

Tight definition and rigid genetic isolation of landraces also bring with them certain risks. As landraces are consolidated and defined in order to assure conservation, their cultural space quickly changes character from their original status as local resources that are taken for granted. The transition from peripheral resource to defined resource usually involves taking the animals from a true landrace subsistence system to a more standardized breed in a production-oriented system. The subtle change in philosophy is not insignificant, for it brings with it the risk that the landrace will become more uniform, more standardized, and that it will leave behind several rare variants as it makes the transition. Production can replace adaptation as a major goal, and this shift in selection pressure can have significant genetic repercussions because adaptation is needed for low maintenance requirements and high survivability.

Landrace conservation has many inherent challenges, not least of which is the very choice of which specific individuals should be included in any landrace. Deciding on where to draw the boundaries around a landrace is not an easy task. Landraces are by their very nature more variable than are other classes of breeds.

Landraces are characterized by biological and adaptational consistency and not necessarily by uniformity of physical appearance. This superficial variability can lead many observers to dismiss landraces as trivial and unimportant when the reality is just the opposite. They are historically and biologically important genetic resources that are adapted to difficult environments. They are generally productive with few inputs, having excelled in survival and adaptation, although generally with less emphasis on high individual levels of production than is typical of the other classes of breeds.

Standardized Breeds

Standardized breeds are the usual populations that come to mind when the word "breed" is used. Standardized breeds are populations of animals that are enrolled into a herd book, flock book, or studbook (these terms are often used for different species, and in this book "herd book" will cover all three). Standardized breeds are, specifically, mated to conform to a written standard (usually called a breed standard, or a standard of perfection) that describes the ideal physical (or in some cases behavioral) type of the breed. The existence of the standard is what gives this group of breeds its name.

Nearly all standardized breeds descend from earlier landrace populations. Exceptions are those breeds that developed from a new combination of previously standardized breeds such as the Columbia sheep breed derived from the Lincoln and Rambouillet breeds. As development and communication increase, breeders get together and decide on certain parameters that are important to the breed. This process defines what is included and what is excluded in a standardized breed. Eventually, in most standardized breeds, breeders decide to "close" the population, which refers to a rule that only offspring of approved parents (generally registered ones) can be registered within the breed and accepted as purebred.

The concepts behind "standardized breed" are the main ones used for most breeds.

As the boundary is drawn around a standardized breed certain characters and traits can easily be left out. The result is that most standardized breeds include more physical uniformity than do most landraces. Some of this uniformity is relatively trivial, such as horned versus polled or coat color. As an example, Welsh cattle in the 1800s were a variable landrace group of adapted cattle. As breeders organized and began to define the cattle as a standardized breed, they selected black (the most common color) as the only accepted color. As a result, the few red, dun, roan, belted, linebacked or white individuals nearly drifted to extinction after being considered outside of the breed definition. These and other variants might well have served important functions of adaptation or production but are now lost to the standardized Welsh Black cattle breed. The key concept is that as a breed moves from landrace to standardized status, breed variation is lost. Effective conservation of both landraces and standardized breeds depends on the recognition of the relative level of variation and the overall philosophy of breed character that each general class of breed brings with it.

Many standardized breeds are internationally important, and standardization is the basic population management strategy that typifies most international breeds. Standardized breeds are genetically isolated by design, in contrast to the landraces, which can be viewed as genetically isolated more by their geographically remote settings (at least historically). The Angus cattle breed, for example, is regenerated by matings of parents that are registered. As a result, Angus cattle from whatever country (United States, Scotland, South Africa, Australia, Argentina) are at least potentially part of the same gene pool. The artificial constraint on matings within the breed allows standardized breeds to function as a single gene pool regardless of how geographically separated the individuals of the breed may be.

As a breed moves from landrace to standardized, variation is lost.

Some standardized breeds do not have a completely closed population, which can lead some observers to conclude that these are not true genetically based breeds. In many situations, though, these still qualify as genetically defined breeds. Quarter Horses, for example, can descend from registered Quarter Horses as well as registered Thoroughbreds under specific circumstances. The herd book is "open," but only in one direction, which preserves the ultimate character of the Quarter Horse. Likewise, the Paint breed allows parents to be Paint, Quarter Horse or Thoroughbred, and the Appaloosa breed allows parents to be Appaloosa, Quarter Horse, Thoroughbred, or occasionally Arabian. The end result of these practices is that such breeds are less rigidly defined than those with completely closed herd books. They are generally consistent enough, though, to be reasonably predictable and therefore do indeed qualify as genetically based breeds. Finally, because they all dip into largely the same pool of genetics they can easily be considered to be a family of breeds of Western Stock Horse type. And, equally importantly, within most of these breeds is a dedicated core of breeders working solely with "foundation" bloodlines. These bloodlines tend to be free from much outside breeding, and can be among the most genetically distinctive animals of the breed.

Tightly closed standardized breeds are typical of most species including cattle, dogs, sheep, and most others. These breeds insist on the registration of only animals that have two registered parents. The result is that the breed is indeed closed genetically, so that nothing can be introduced. This protects the integrity of the breed's genetic package, although it can also be damaging if the package becomes too narrowly constrained such that viability and overall vigor begin to suffer. The negative consequences of rigid closure are beginning to manifest themselves in some horse, cattle, and dog breeds, and the future of the strategy of tightly closing standardized breeds is now more open to debate than it has been for the last century. Standardized breeds are a relatively recent development (about 200 years old, as contrasted to the backdrop of 10,000 years of agriculture), and strategies for their long term genetic management may still not be fully developed or understood.

Modern "Type" Breeds

In addition to landrace and standardized breeds is a category that is more open than is true of genetically based breeds. These include many of the Warmblood horse breeds, the Pinto horse, and a few other breeds in a number of species.

Modern "Type" breeds can be productive, but do not serve well as genetic resources.

In these "type" breeds, character is based more on external type or performance than it is on genetic continuity with any specific gene pool. The result is some very useful animal populations that have only limited genetic consistency and therefore limited predictability in reproducing a specific type. These breeds can be productive, but serve very little utility as genetic resources. They are therefore outside the realm of genetic resource conservation.

Breeding practices further confound the inclusion of many Warmblood horse breeds into the overall class of modern type breeds. Within many of these breeds is a core of horses that are indeed "purebred" in the genetic sense of having ancestors only of that breed. While these are a small minority in most Warmblood breeds, they reflect the conservatism of traditional breeders and their allegiance to local types and breeding. This core within each of several internationally popular Warmblood breeds conserves the unique portion of what is becoming an increasingly homogenized group of breed resources. Unfortunately, this core and the more numerous outbred portion of the breed usually are known by the same breed name, and so an accurate appreciation of the breed's status can be nearly impossible to ascertain without detailed and prolonged study. This pattern is repeated throughout a few other species, most notably several poultry breeds.

Industrial Strains

Industrial strains are usually not characterized as breeds. In most cases these are very tightly selected lines within a breed, or hybrids based on a few breeds. The result has been exquisitely productive animals that are highly selected for a specific production niche that requires advanced nutritional, environmental, and housing support. Industrial strains are scientifically selected, mated and documented. Industrial strains are successful because they are indeed productive.

Industrial strains are productive in highly controlled environments.

Industrial strains are not documented in herd books because they are owned and managed by multinational corporations and are available only through corporate sources. Individual private breeders are not involved, so the documentation of registrations and the like has become superfluous for industrial strains. Most broiler and egg-laying chickens, turkeys, and industrial swine are now industrial strains rather than standardized breeds. Dairy cattle could also be considered in this class, although these are more widely held and reproduced by private individuals. Within dairy cattle breeding the role of the semen companies serves as a bottleneck through which genetic variation is managed and largely minimized, and for dairy cattle breeding these companies play the role that the multinational corporations have for other species. This is due to nearly all dairy cattle being reproduced by artificial insemination so that the decisions of semen companies influence entire breeds. The practical result of semen company influence has been a trend toward increasing relatedness among most cattle of any of the dairy breeds. In addition, like other businesses, these companies are merging to consolidate the control and marketing of genetic

Industrial broilers are very uniform both genetically and phenotypically. Photo by D. E. Bixby.

material into fewer and fewer hands.

The control of industrial strains (especially poultry and swine) by large corporations assures that small private breeders have no role to play. They are interesting populations because variability is low and predictability high. Industrial strains can and do become endangered or extinct through corporate decisions (usually after mergers between multinational corporations). Very little or nothing can be realistically accomplished to reverse that trend, as private breeders simply do not have the resources nor the infrastructure necessary to duplicate the industrial selection environment that maintains these strains.

Feral Populations

Feral animals are domesticated animals that have escaped domesticated settings and have returned to a free-living state. It is a peculiar fact of biology that the truly wild type and wild genetic strain are never fully regained by feral populations, even though some feral animals do indeed approach the wild ancestral type of the species.

Feral animals are interesting due to their being returned to a selection environment where nature, rather than humans, decides which ones reproduce and which ones succumb. Distinctive feral populations usually arise from a few founders and then experience long isolated periods of survival under natural selection. Such populations have much in common with landraces, although nature rather than humans imposes all selection. Most feral populations, however, have broad genetic variation due to constant infusion of new recruits from a wide variety of genetic sources. These populations are unimportant as genetic resources or as targets of conservation. Only a very few feral popu-

Only a few feral populations qualify as genetic breeds.

Ossabaw Hogs are a unique long-term feral breed from a barrier island off of the coast of Georgia. Photo by D. P. Sponenberg.

lations qualify as genetically distinct breeds with limited variability. The few that do qualify are indeed fascinating for their adaptation and survival traits. Examples of these occur for all species: Colonial Spanish Horses (specifically Spanish Mustangs), Island populations of goats and sheep, and swine from a constrained (generally Iberian) origin.

Conservation of genetically distinct feral livestock populations that qualify as breeds is problematic. Most of the populations of conservation interest come from relatively isolated regions. On many islands the feral livestock are endangering native flora and fauna, and the usual decision is to favor conservation of the natural resource rather than the introduced one. Conservation of the lineage of feral animals can be accomplished out of the original environment, but brings with it the very real conundrum that the selection environment is new and changed from the original so the genetic result is different than would have been the case in the original feral state.

A Review of Classes of Breeds

The various classes of breeds have profoundly different histories. This is due to the specific combination of founder effects, relative level of genetic isolation, intensity of human selection, and physical environment in which they function. As a general rule landraces are more variable than standardized breeds, although there are some exceptions. Industrial strains are the least variable of all. Feral populations include some that are highly variable, and others that are among the most genetically homogeneous of any breed populations. The modern type-based breeds are usually fairly genetically variable. It is important to understand the relative level of genetic variability of these different classes of breeds, as well as the historical and cultural reasons behind these different levels. Appreciating the genetic and cultural background of breeds is essential if organizational and breeding strategies are to be successful in conserving them.

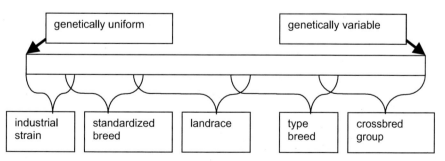

Classes of breeds vary in their relative level of genetic uniformity.

What Should Be Included in a Breed

The question of what to include within a breed population becomes an interesting and thorny problem for many breeders and their livestock. This problem plagues rare breeds of conservation interest much more than it does mainstream production agricultural breeds, because rarity implies the need to include every individual that is truly of the breed. The goal of breed conservation should always be to try to eliminate all animals that are not of the breed, and to include all of those that are. Decisions are thereby pulled in opposing directions to avoid leaving anything out, while also avoiding inclusion of extraneous and nontypical animals.

The goal of breed conservation is to include all animals that are truly of the breed, and to exclude all that are not of the breed.

Determining the boundaries of a breed, especially a landrace breed, can be a challenging and complicated task. Reflecting on the identity and character of breeds can greatly aid this process. Breeds are repeatable genetic packages. However, they are also more than this as they have all been developed within the context of a specific human culture and use. Combining the genetic character and cultural setting of breeds helps to direct the decisions of what to include in a breed population.

The three most useful investigations into animals presented as members of a breed include phenotype, history, and finally genetic analysis. Any of the three, if used alone, can lead to a wrong conclusion. In contrast, when all three are used in concert it is rare to misclassify animals.

Phenotype is important because all members of a breed should consistently reflect breed type and conformation, apart from certain superficial cosmetic characteristics. The external indicators of breed membership are not trivial, for they mirror the underlying genetic package and are therefore a very good and easily observed indication of relative congruence with what is expected of members of the breed. Phenotype can be misleading in the case of individual animals, but is much less likely to be so when considering entire populations. So, an individual horse presented as a Colonial Spanish Horse is much more difficult to assess than is a herd of 100 individuals. This is true because the 100 individuals are much more likely to betray any deviation from the breed package than is a single selected individual. Investigation of the phenotype of candidate animals and populations is usually the most economical and easy of all investigations, and so is generally used first as a quick route to a decision concerning breed membership.

Investigation into the history of candidate animals or populations is usually

Colonial Spanish Horse populations, such as these Pryor Mountain horses, are usually included in conservation work as a result of qualifying on the basis of complete investigations into phenotype, history, and genetic analysis. Photo by D. P. Sponenberg.

accomplished after the phenotypic investigation – or sometimes concurrently with it. History is important because most breeds spring from a limited area or a limited influence. Colonial Spanish horses (Spanish Mustangs and Barbs) provide a good example here, as feral herds with appropriate phenotype are continuing to come to light. The history of these must be evaluated as to the known or suspected introductions of outside horses. Historical investigation can be quite difficult because many of the people involved have a vested interest in the outcome of the process. Those who favor inclusion of the population into the breed will often indicate on a pristine history of absolute purity. Those who would detract from it will point to long and consistent inclusion of outside influences. Sorting through these different threads can be quite difficult, although weighing the history against the external phenotype can greatly aid in coming to an accurate conclusion. A history of constant introduction is simply not consistent with a resulting uniform population, and neither does a history of isolation lead to a highly variable population.

The three pillars of breed investigation are phenotype, history, and genetic analysis.

People familiar with the candidate population and its type can evaluate phenotype, even if somewhat subjectively. This may not satisfy all challenges to scientific accuracy, but is frequently all that can be accomplished for large, extensively raised populations such as feral horses or range cattle. An approach

This Choctaw strain Colonial Spanish horse has a strongly typical head, accurately betraying his breed identity. Photo by D. P. Sponenberg.

to doing this has been undertaken with Colonial Spanish horses, with the development of a score sheet to focus on important physical characteristics that are typical of this type of horse. This score sheet is presented in Appendix 1.

An alternative to a subjective analysis is a linear measurement assessment. In this type of assessment very specific distances are measured on a number of animals in the population. These measurements, when analyzed in their totality, are remarkably accurate in pinpointing membership in the breed, and are likewise good at eliminating nonconforming animals. The sorts of measurements vary with the different species, but usually include: height at withers, height at top of croup, height in middle of back, halfway from dorsum to belly, width of croup, anterior width of croup, posterior width of croup, heart girth, width of head, length of head, length of face, distance between eyes, circumference of muzzle, width shoulder to shoulder, length of body, depth of body, and top line poll to rump.

Wherever possible, animals and populations that pass muster from the phenotypic and historical aspect should also be investigated genetically. For most cattle and horse populations it has been possible to bloodtype candidate animals as a final step in deciding on breed membership. Recent advances in DNA fingerprinting have allowed this technique to largely supplant bloodtyping as an investigative tool. Unfortunately, bloodtyping and DNA fingerprinting provide different information to investigators, and these are not interchangeable. DNA fingerprinting has greater accuracy for parentage verification, and this is one reason it has supplanted bloodtyping. The strength of bloodtyping, though, is the huge repository of information acquired over a long time that allows breed-to-breed comparisons to be made. As a result, bloodtyping has very strong advantages over DNA fingerprinting for most breed determinations. This is, however, changing as more and more DNA results are amassed. Also, DNA fingerprinting and bloodtyping are common only for horses, cattle, and alpacas. This is an additional challenge when investigating other species, so that for several species the decisions rest more on phenotypic and historical investigations.

An example of the value of employing all three approaches can be seen in many breeds, but especially in Colonial Spanish Horses. One population, the Nokota horses, have a history of descent from Chief Sitting Bull's herds. From there they went to the Teddy Roosevelt National Park, and the story becomes clouded with the possibility of crossbreeding on some parts of the range. The horses look variable as to type, but the most typical individuals are very good representations of the Colonial Spanish Horse type. It was important to decide between these as remnants of a historically important population, or as odd re-segregants from a crossbred population. Unfortunately the DNA and bloodtyping results clearly pointed to recent crossbreeding across all of the types within the population.

Other populations yield still different illustrations of the importance of all three evaluations. Wild horses on Jarita Mesa in New Mexico are from an isolated region. A phenotypic evaluation showed the horses to be very uniform, but also very slightly "off type" (somewhat short, somewhat short-headed) for the usual Colonial Spanish horse herd. An evaluation of the history pointed to the release of a herd of Welsh Ponies into the area in the 1930s drought. In addition, the DNA and bloodtyping evidence pointed to the same conclusion. The intervening years of maintenance as a closed population had resulted in a very uniform population of very handsome horses – although with nothing to offer to the conservation of the Colonial Spanish Horse.

One Breed or Two?

Animal populations within any species can be characterized historically, genetically, and politically. The results can be viewed as a branching tree, where recent divisions or splits form the outermost branches, and earlier divergence is a major branch or a split trunk. For example, the difference between humped and humpless cattle is a major branch low down on the tree of cattle breed relationships. This is in contrast to the difference between Angus (black) and Red Angus cattle, which split relatively recently.

For conservation it is wise to link populations that are more like one another than any other breed.

Splitting and branching continue within breeds, and if the relationships are taken below the breed level the branches are usually referred to as strains and varieties. These are all generally considered to be parts within the single breed, but to be distinct

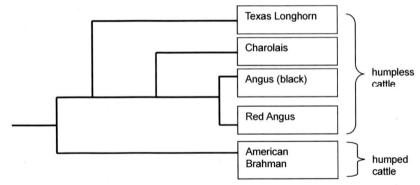

Breeds branch from each other sequentially, and the horizontal distances between branches reveal the relative relatedness of different breeds.

enough by history, genetics, or phenotype to warrant some additional identification. For example, poultry varieties within a breed vary by color, plumage, or comb type.

The mental picture of a tree of breed groups, breeds, varieties, and strains is useful in pointing out a very real and confusing issue in breed definition. At what point along a branch should populations be considered distinct for maintenance and conservation purposes? This question has no easy answer, because the answer is rooted both in biology and in the political environment surrounding breeds. One useful strategy is to link populations that are more like one another than any other breed. That strategy may well group close cousin breeds that have been separated for only a short while. Any decisions to group together or split apart should be based on the reality of breed persistence and viability instead of on politics or personal preferences.

Pineywoods cattle can serve as an example of the difficulties encountered in deciding to group or to split. Pineywoods cattle persisted in the hands of a very few dedicated breeders whose numbers dwindled in the last half of the 1900s. As a result of fewer and fewer breeders, the herds became increasingly isolated from one another both geographically and genetically. By the late 1900s the few remaining herds had histories of complete isolation for several decades, and in most instances for a century or more. In one sense these could each have been considered separate breeds, and each could have been managed to maintain the isolation. However, the resulting lines would each have been critically endangered, and all of them in peril of extinction directly from their low numbers which make inbreeding depression unavoidable.

An alternative view of the Pineywoods situation is that each of the strains had a historic link to the same earlier population of Spanish-based humpless cattle in the southeast. That origin had led to a similar phenotype across all the strains, and a phenotype that was unlikely to be confused with later zebu (humped) or northern European (nonhumped) cattle. By grouping the strains together in a single breed it became easier to manage the population for viability into the distant future, while at the same time not endangering the distinctive characteristics of the cattle.

The answer to "at what level to group" is never simple. In some situations closely related breeds are kept separate by registration, such as with Dales and Fell ponies, or Clydesdale and Shire draft horses. In other situations, history and phenotype put considerable variation under a single landrace breed group, such as the Colonial Spanish Horses with strains as geographically diverse as Marsh Tackies, Florida Cracker, Pryor Mountain, Choctaw, and Doroteo Baca New Mexico horses.

The Pineywoods breed is an example of grouping with some recognition of the distinct strains within the breed. They are grouped in order to conserve them in one population. The Dales Pony and Fell Pony, in contrast are maintained as separate populations by their registries even though they are very closely related.

Breeds as Gene Pools: Variability and Predictability

Variability and predictability can be viewed as extreme endpoints along a single line. At one end are populations that are completely uniform genetically. These are very predictable. At the other extreme are populations (almost always crossbred) that are extremely variable and therefore unpredictable as to what they

The Agricola strain of Pineywoods cattle is one of several distinctive strains of the breed. Photo by D. P. Sponenberg.

26 *Biology of Breeds*

Fell ponies are useful multipurpose British ponies, closely related to Dales ponies. These two breeds form a useful pair of breeds that are genotypically, phenotypically, and historically linked. Photo by J.K. Wiersema.

will produce when mated. The point at which populations are useful as predictable genetic packages (breeds) is closer to the genetically uniform end than it is to the genetically variable end. The specific point occupied by a breed along the continuum varies breed to breed, as was discussed in the previous chapter.

Breeds serve as genetic resources because they have predictable combinations of genes throughout most or all individuals of the breed. Predictability is vitally important to breeds, and this implies a certain level of genetic consistency. Populations that are highly genetically variable are no longer predictable, and predictability of performance and appearance is the very essence of the importance of breeds. Owners choose specific breeds because they are interested in a certain appearance, performance, and behavior. Predictability allows owners to be satisfied with their breed choice.

For breeds to function as genetic resources they need variability for health and uniformity for predictability.

Added to the need for predictability is the need for breeds to be maintained as viable genetic entities so that they can survive to serve as genetic resources. The need for viability implies some level of variability because truly homogeneous populations are not self-sustaining due to losses in reproductive health and overall viability. Populations begin to lose vigor and reproductive fitness if genetic variability becomes too low. For breeds to function as viable populations as well as genetic resources, they need to have enough variability to be healthy while at the same time having enough consistency to be predictable. Only by balancing these two somewhat opposing forces is it possible to assure that breeds can serve as viable genetic resources

Genetic Organization of Breeds

Breeds are organized with different structures that have consequences for their management as genetic resources. A few different patterns of structure are common to most breeds. Understanding the manner in which a breed is organized genetically is essential if breed management and conservation are to be effective. The genetic organization of nearly every breed is a reflection of its past history of management and selection.

The most common organizational structure for standardized breeds is a tiered structure, which resembles a skyscraper. The bottom, small, portion is called the nucleus, and comprises elite herds that contain the most desired animals in the breed. The entire breed rests upon this genetic base. Next up is a larger layer comprising multiplier herds, of somewhat more average quality but still typical representatives of the breed. The top and largest tier is the commercial segment, consisting of herds that are providing animals for final products instead of specifically for purebred breeding replacement.

The tiers within a standardized breed tower usually interact as a "closed nucleus" system. This refers to the general flow of genetic material within the breed. The nucleus herds generally produce their own replacements with no introductions from the other levels. This results in little or no gene flow into this portion of the breed from the rest of the breed, but with outflow of genes (via

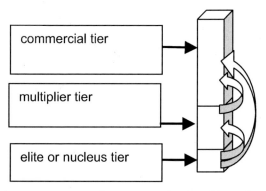

The genetic organization of most standardized breeds is a closed nucleus system.

breeding animals) from this group to the rest of the breed. The multiplier tier generally buys in replacement stock (especially males) from the bottom elite tier, and also saves its own replacement stock (some males, and mostly females). The commercial tier usually buys males from the multiplier tier or from the elite nucleus. The overall pattern is a flow of genetic material from the bottom to the top of the breed tower, but little to no downward flow to affect the foundational base. Closed nucleus organization is very typical of standardized modern production breeds. This structure organizes the genetics of the breed to have a relatively small base (the elite herds) under a tall structure (the overall population).

In industrialized breeds the organization becomes even more extreme because the elite tier is entirely closed, and is multiplied to produce huge numbers of animals at the multiplier and commercial tiers. The population may be very large, but the genetic base is small. This provides for the genetic uniformity and predictability that are so desired in commercial stocks, but also constrains the population to a very small genetic foundation.

Industrial stocks have a narrow genetic base.

The most extreme example of this structure is the industrial egg laying and broiler chickens. The vary narrow base in these bird populations consists of four closely held foundation lines for each industrial strain. A specific two of these are crossed for a maternal multiplier line, and the other two for the paternal side of the multiplier step. These multiplier populations are larger than the original four founding lines, and produce the vast numbers of birds that are used for final commercial production. Holstein cattle, while not as extreme an example, are also similarly based on a narrow foundation. The genetic consolidation in Holsteins is largely a consequence of decisions made by semen companies, and a great deal of genetic narrowing occurs through the use of paternal half sisters as the dams of widely used bulls.

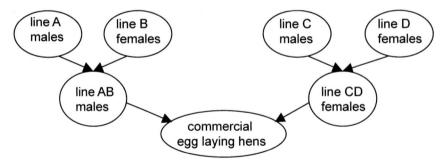

Genetic production of industrial egg producing hens depends on an intricate system of managing genetic lines and their breeding.

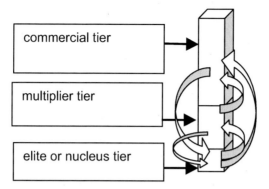

An open nucleus system allows genetic material to flow back into the elite or nucleus tier through the use of exceptional animals from the multiplier and commercial tiers.

An alternative organizational model, rarely used, is the "open nucleus" system in which males from the elite nucleus are used at both the multiplier and commercial tiers (similar to the closed nucleus model). Males from the multiplier tier are used at their own tier as well as the commercial tier (also similar to the closed nucleus model). Importantly, though, a small percentage of females with superior performance move down the structure from the commercial and the multiplier tiers to the elite nucleus tier. This seems to be a subtle and inconsequential shift in strategy, but provides for much more rapid dissemination of superior genetic combinations throughout the breed than does the closed nucleus approach, because superior animals in the upper tiers have an opportunity to have a wider influence on the breed than is true in the closed nucleus model. This approach broadens the genetic base of the breed.

Merino sheep breeders in Australia have benefited from using open nucleus breeding schemes. Photo by Scott Dolling.

The open nucleus system has been rarely used, but has been adopted by some sheep breeding cooperatives in Australia. Each cooperating breeder contributes his elite ewes to a group flock, from which all the members can take rams for use back in their own flocks.

The open nucleus system is also being promoted as a useful strategy for village-based animal breeding in developing countries, as it widens participation and supply in programs that target improved production at the local level.

Landraces usually have a genetic organization very distinct from that of standardized breeds. Rather than a tall skyscraper, these breeds are much more like a short one- or two-story building with many separate rooms. Each subpopulation (strain or line) within these breeds is variably isolated from the others. In many cases, breeds organized in this manner comprise several herds that each persist in complete genetic isolation from one another. The genetic similarities in these breeds come from long-standing founder effects and from consistency in the selection environment, and not from more recent genetic transfers among the strains. In addition to the foundational strains of landraces there are composites built from multiple foundational strains, which are like a second-story room over several of the first-story rooms, supported by the genetics of the foundation strains.

Genetic similarities in landraces come from founder effects and a consistent selection environment.

Pineywoods Cattle are an example of a "low, broad building" breed. Many family herds of Pineywoods cattle have been totally isolated from one another since the early 1900s. These family herds are excellent examples of genetic bloodlines. In addition to these foundation herds are more recently established herds that are based on a composite of varying combinations of these foundation herds. The overall gene flow is from foundation herds to composite herds, as well as from one composite herd to another. The foundation herds, however, function in nearly total isolation from one another, so that the breed has several distinct and important isolated reservoirs of genetic material rather than a single

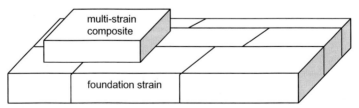

The genetic organization of most landraces resembles a short one-story building made up of distinct foundation strains, with a second story of composites built from the foundation strains.

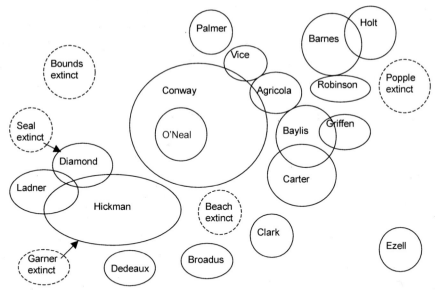

Pineywoods cattle strain relationships. Overlapping ovals indicate that strains are related through past exchanges of breeding animals.

core. The isolated strains do have a high risk of loss or extinction due to dispersal or natural disaster.

The persistence of distinct foundation strains assured that genetic diversity is maintained within the breed. Keeping these separate groups intact assures that distinct lines are present within the breed, so that every animal within the breed has an unrelated outcross in the event that such measures are needed. The usual reason for needing such an outcross would be diminished vigor in a line as a result of inbreeding.

Persistence of strains helps rare breeds to maintain genetic diversity.

The importance of distinct bloodlines is well illustrated by the recent fortunes of the Java chicken breed. Javas are, by their standard, large robust birds. Years of declining populations led to isolation of a few pockets of birds of the breed, and each began to undergo loss of vigor and breed character. This decline was largely through inbreeding depression, and further sealed the fate of this breed as a somewhat trivial relic from the past. Fortunately Peter Malmberg, of the Garfield Farm Museum in Illinois, was able to assemble birds from the remaining few strains and to cross them. The resulting birds benefited from the hybrid vigor of the cross, and admirably regained both vigor and breed character.

At this point, though, it is important to consider next steps. One strategy is to

Within the Pineywoods breed are several unique strains, such as this Holt cow. Photo by D. P. Sponenberg.

continue to work with the composite population which resulted from the cross, but this is likely to again undergo decline as the interrelatedness of birds increases with no possible outcrosses. A second strategy is to closely monitor the original, subpar, strains and maintain these as reservoirs of genetic variability needed for the future. A third option is to carefully manage the original strains by adding in about one quarter or one eighth breeding of the other strain in order to assess overall viability and breed character (hoping to improve both) while still maintaining important genetic differences in the strains. This is likely the best choice because it is practical and not likely to fail. A fourth choice, that could succeed as well as the third strategy, is to combine the strains, then quickly subdivide the resultant crossed population into several different locations and breeding populations in order that new strain differences emerge. The transfer of breeding animals among such populations needs to be carefully monitored in order to assure that the entire population does not go in a single direction with one genetic strain influencing all others. The challenge is to maintain breed type, breed character, and also vitality and genetic strength that are greatly influenced by the existence of relatively unrelated bloodlines within a breed.

The different organizational patterns of breeds have profound consequences for conservation and genetic management. Standardized breeds with a very strict closed nucleus structure are, at least superficially, the easiest to conserve,

Biology of Breeds 33

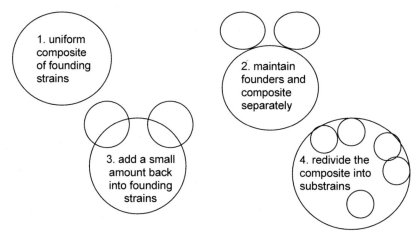

Different strategies for managing the genetic diversity in Java chicken populations.

because a reasonable sample of the elite tier is effective in saving most of the genetic variation in the breed. At the other extreme, the "low, broad building" organizational structure is more difficult to conserve, because each and every separate strain should be located and sampled in order to save the entire genetic diversity of the breed. It is a common misconception in breed conservation programs to assume that the closed nucleus structure is typical of all breeds. The result has been that many important founder strains of landrace populations have been overlooked in conservation work, and many have become extinct.

Gulf Coast sheep have a similar history to that of Pineywoods cattle, but the organizational structure has recently tended more towards a standardized

Java chickens are one breed to benefit from the boost provided by genetically distinct foundation strains. Photo by Larry Reynolds of dogpix.com

Patterns of genetic organization of breeds have consequences for conservation.

breed than has the Pineywoods structure. Some of the distinctive bloodlines that once existed have been lost in the composite of the newly defined population. In addition, the resulting composite is increasingly based on relatively few founding strains rather than a more balanced representation of the variation in the overall landrace. Fortunately, a few isolated and unique strains persist, and the future genetic health of the breed will depend heavily on breeder decisions regarding these strains and the variability they represent.

Bloodlines within a Breed

Bloodlines or strains within a breed are those subpopulations that have been isolated from one another for several generations (usually four or more) with the consequence that they are somewhat genetically distinct from the other bloodlines. Bloodlines are usually linked to certain breeders or farms, and can be distinct historically and genetically. The degree of distinction of bloodlines can be problematic, for it is easily possible to come back to the question of "what is a breed?" and the specific degree of genetic distinction that should be included in or excluded from a single breed.

Some bloodlines could indeed be considered to be their own breeds, although such an approach would vastly multiply the number of breeds that need conservation help. The approach that the American Livestock Breeds Conservancy has taken is to consider populations as bloodlines (rather than breeds) if they are more like one another than they are like any other breed. This strategy is reflected in the Pineywoods cattle, which are considered one breed of several distinct

Gulf Coast sheep still maintain isolated foundation strains. Photo by D. P. Sponenberg.

bloodlines, rather than a group of cousin breeds. The bloodlines are reasonably distinct historically, phenotypically, and genetically, yet are much more like one another than they are like any other breed. Even though some of the founding strains have been genetically isolated for more than 100 years, their similarities one to another support their inclusion under a single breed identity.

The bloodlines within a breed can be very important reservoirs of genetic variation, and managing these within the overall breed is important to long term breed survival. Bloodline conservation can lead to fads and shifts in popularity of one bloodline versus another. Some of these shifting fates may be related to production potential or to breed character, in which case these shifts may be warranted. Many times, though, the popularity of a bloodline is related to the advertising or show ring success of a capable promoter. In those cases a bloodline can easily swamp a breed with little underlying genetic reason for doing so. In nearly all cases it makes sense for a breed association to work to effectively conserve all of the component bloodlines of a breed.

> *Bloodlines within a breed are reservoirs of genetic variation.*

In the Texas Longhorn breed several of the eight foundation lines have either become extinct or have been crossed so extensively with other lines that they have lost their original genetic distinctiveness. This was done as a response to shifting emphasis on certain phenotypic characters that were present more in some lines than in others, as well as an increased tendency to use semen from certain popular sires. The result is a serviceable breed, but one that is much changed from the original.

Breed Histories

Understanding breed history is essential if breeders are to adequately steward the breed as a genetic resource. Each breed has been shaped into its current form by a unique history. Breed histories must be clear and accurate to the extent possible. The reason that clarity is essential is that most breed origins are obscure and shrouded in mystery. As a result the claim is all too common for a breed to have originated in the mists of time and that it is "one of the oldest pure breeds in existence." This is clearly not true of most breeds, and understanding the relationships of the breed, its foundation influences, its early and present environment, and its history of selection pressures can all help to guide management and conservation of the breed as a genetic resource.

Breed history should accurately reflect the source of founders and the way in which they were used to contribute to the current breed. Were outside influences common or rare? How did selection produce the animals that survive today? Each of these aspects is key in forging a breed as a genetic resource, and each should also influence decisions for breed maintenance and conservation.

Breed histories can be complicated. Most involve a reasonably well-documented early foundation. Then, in the case of many rare breeds, a period ensues in which very little is documented or known. If the breed had remained popular then it is obvious that people would have taken more interest in documenting what was happening. A breed often slips into rarity because people quit caring about its fate, with the result that many unique and useful rare breeds have lapses in breed history.

Examples of breed histories with gaps in the middle are the Colonial Spanish horses and the Florida Cracker cattle. In both cases an exciting amount of detail is known about the founding events, though centuries ago. Details also tend to be known about the expansion of these breeds to become numerous in the 1800s. Then a silence falls on documentation for a century until recent conservation efforts began. The conservationists are only able to focus on a few, traditional family lines, and it is from these that the present-day breeds descend. Breeds such as the Texas Longhorn do indeed go back to the millions of such cattle on the plains of Texas in 1850, but do so only through a very limited number of family lines, each of which has become a bottleneck. The same is true of Colonial Spanish Horses and Florida Cracker cattle. Understanding history can help breeders to manage and conserve these breeds more effectively, as their histories are very different from one in which a wider sampling of the previous millions has persisted to the present.

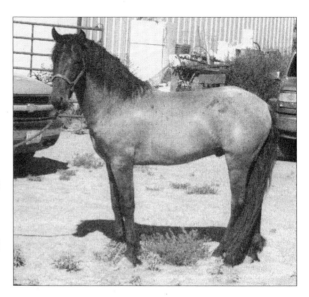

The Mount Taylor, New Mexico, strain of Colonial Spanish horses is one of many strains that connect this breed to its rich past. Photo by D. P. Sponenberg

Tennessee Myotonic goats are one breed with a very profound link to a specific geographic region. Photo by D. P. Sponenberg.

Geography and Source Herds

Geography can play a large part in identifying source herds for many breeds. This is especially true for landrace breeds that present complicated conservation challenges. The original geographic region of a landrace is especially likely to provide newly discovered herds, while regions outside of the geographic region are very unlikely discovery sites.

The Tennessee Myotonic goat can serve as a useful example. In middle Tennessee many traditionally managed herds of these goats survive. Their owners persist outside of the formal structure of breed associations and registries. A pure type Tennessee Myotonic goat is much more likely to come from this region than it is from northern Idaho – especially when considering historic documentation of the herd's foundation and other details of its lineage.

The link of geography and landraces is essential to their status as genetic resources forged for adaptation in specific environments. Taking geography into account when considering "found" or newly discovered herds or individuals is therefore appropriate and helpful in recruiting these animals into conservation programs. The link of geography and genetic resources also can point to regions where active searches are needed to identify remaining pockets of rare landraces so that they can be brought into effective conservation programs. An example of success in a geographic approach to conservation work is the discovery (or rediscovery) of the Baca and Mount Taylor strains of Colonial Spanish horses in New Mexico – right where the hub of the diffusion of this important North American breed occurred.

The link of geography and landraces is essential to their status as adapted genetic resources.

History, geography, and genetics usually all go hand in hand for effective breed definition and breed conservation. All avenues should be thoroughly explored in order to ferret out what is most likely to be true, and most likely to result in meaningful conservation of legitimate genetic resources.

3. Breed Standard

A good breed standard is an essential tool that greatly helps breeders to maintain and manage a breed as a genetic resource. The standard should be the mental picture that drives selection decisions by all breeders. The concept of breed standard can be a tricky one for many breeds, and different classes of breeds need different sorts of standards.

Standardized breeds, logically, have a standard. The standard is indeed integral to the overall culture of breeding these standardized breeds, and is nearly an assumed underpinning of much of the breed community. Standards for most standardized breeds are "prescriptive." The standards for these breeds prescribe what the ideal animal should be. This ideal may or may not actually exist – it is a target for all breeders to aim for. An example is the breed standard for Leicester Longwool sheep, a standardized English breed developed in the eighteenth century by Robert Bakewell as one of the first deliberate attempts at breed formation. The standard prescribes a very specific combination of head shape, head carriage, color, leg conformation and placement, body size and style, and fleece characteristics. By the standard, the ideal Leicester Longwool sheep can be seen as a single sheep, and this is the guide for breeders' selection decisions as they aim for that target of the perfect Leicester Longwool sheep.

The breed standard drives selection decisions.

In contrast, the best standards for landraces are "descriptive." These describe, rather than prescribe, what the animals actually are. This is subtly different from a prescriptive standard, but understanding this difference is essential if landraces are to be successfully conserved. Making a descriptive standard is more difficult than a prescriptive one, especially as most breeders are interested in some degree of uniformity and superiority. These goals can tug against the variability inherent in landraces. An example of descriptive standards are those used by the Spanish Mustang Registry, Horse of the Americas, and Southwest Spanish Mustang Association for the Colonial Spanish horse. These give a range of types varying from a heavier, generally northern, type and a lighter, generally

Leicester Longwool sheep are one of the oldest standardized breeds. Photo by D. P. Sponenberg.

southern, type, with a range of variation between these extremes. The Colonial Spanish horse score sheet for type includes acceptable variation in such details as head shape. A range, rather than a specific single type, is appropriate for this landrace breed. The ideal Colonial Spanish horse is not a single horse, but instead fits within a range of acceptable types.

A potential problem with landrace standards is that including the more divergent foundation strains of the breed under a single breed umbrella does not always sit well with the very folks that saved the divergent strains in the first place. This can seem something of a paradox, as these very people should be most interested in the conservation of the breed. Breeders of the various strains are very likely to consider their own strain as the only true representative of the breed. They can, and sometimes do, consider that inclusion of strains other than their own family strain diminishes rather than enhances conservation. Some traditional breeders can refuse to participate in a group effort that includes more than their specific strain. In order to succeed as effective conservation tools, the standards for landraces must address these issues and must be clear as they address the issue of acceptable ranges of variation. The standard must be specific as to allowable variation, or it can become so vague as to be overly inclusive and not useful for weeding out variation that is not within the limits of the landrace.

Standards for standardized breeds are prescriptive. Standards for landrace breeds should be descriptive.

The challenges of a prescriptive breed standard are well illustrated by the historic standards for some chicken breeds. Some poultry breeds have long had standards in which the color prescriptions for males and females were unable to be met by a single interbreeding population. As a result, show ring breeders would have "male lines" to produce show males, and would match these with females from "female lines" in order to put winning groups together at shows. The two lines were never crossed, because the females of the male line and the males of the female line failed to meet breed standards. These off-standard birds were essential to produce the show winners of the opposite sex. Experienced breeders realized that this prescriptive approach did not reflect the reality in the breeding yard, because single populations could not accomplish the prescribed goal of the standard. A more descriptive standard would have allowed for the males to be appropriate for females of the same breeding line, and some breeds of poultry have now changed their standards to reflect the reality of the type within a single true breeding population.

Landraces and local breeds existed before the standards that describe them.

It is important to realize that most landraces and other local breeds existed before a breed standard was derived for them. This is an essential detail if important variants are not to be lost for future generations. Standards for landraces must allow more variation than standards for standardized breeds. Most landraces eventually move towards the philosophy of a standardized breed. They lose much in the transition, however, so designers of standards for landraces should

Colonial Spanish horses vary somewhat as a reflection of their variable foundation herds. Photo by Marye Ann Thompson.

be especially diligent in trying to be specific enough to assure their unique status as important genetic resources while not being too strict so as to exclude legitimate variation.

The Development of a Breed Standard

Breed standards range from very narrowly defined to more broadly defined. It is common for standards to be so broad and vague that they could indeed describe just about any well-conformed animal. In order to be truly useful, breed standards should be specific and should result in a description of the breed that could not be confused with descriptions of other breeds. The combination of conformational peculiarities unique to a specific breed constitutes "breed type" and sets one breed recognizably apart from others. Breed type is the sum of physical, behavioral, and functional traits, and while these may not always relate directly to functional conformation, they do serve as important indications of the integrity of a breed's genetic package.

Breed type sets one breed apart from others.

Breed standards all too commonly indicate only how animals should be conformed. This usually relates to overall soundness, and descriptors such as "level," "broad," "well made," and "strong" usually do little to convey the uniqueness or character of any animal or breed. The more precise and breed-specific the standard is the better it serves to guide breeders in maintaining the breed.

It is common for breed standards to emphasize external and easily observed characters such as color, and to skirt discussion and description of breed-specific traits. These are generally difficult to describe because they are complicated, and it is difficult to

African geese have a unique breed type that makes them difficult to confuse with any other breed. Photo by D. P. Sponenberg.

Breed Standard 43

Rocky Mountain horses have unique gaits that help to define the breed. Photo by D. P. Sponenberg.

marshall the vocabulary that can convey the subtleties of breed type. An example is a description for the shape of the head. It is in the arena of type traits that a breed standard is most useful if it can be written to truly convey the uniqueness of the breed type. A possible description of a sheep head is "broad muzzle with a straight profile, with lower teeth meeting the upper dental pad evenly. Ears are fine and moderately large, and carried horizontally from the head. The jaw is deep, wide, and full. Skin is dark with hair usually white, producing a blue appearance. Long, lustrous wool grows from the topknot but cheeks lack wool with no tendency towards woolblindness." This head description fits Leicester Longwool sheep and very few others. Those others could easily be eliminated by description of the breed standard for other body regions. The overall result of descriptions like this is a breed-specific document that can help to guide breeders.

A more mathematical approach to breed documentation is that of linear assessment, which compiles several specific distances (eye to eye, shoulder to hip, many, many others). In most situations these measures are compiled from several animals and then are given as an average and a range of values. These measures, when combined, fairly accurately reveal breed specific conformation, although they lack the ability to instill a mental picture that the classic breed standard allows.

Functional traits have equal importance to physical traits for breed function and integrity. Standards can include mothering ability, foraging ability, parasite resistance, ease of parturition, fertility, breeding season, and longevity. In some horse breeds gait is an essential component of the breed standard. These functional traits are difficult or impossible to assess by visual inspection, so most breed standards do not include them.

Breed Standards and Genetic Diversity

The goal of any breed standard is to help breeders visualize characters and traits that should be included as typical of a breed. The standard guides observers (especially breeders and judges) as to what should be ideal for the breed, what is marginal for the breed, and what is outside of breed parameters.

The level of specificity of a breed standard varies from breed to breed. For standardized breeds it is common to have a fairly narrow breed standard that is based on an ideal that may or may not have been achieved – or may not even be achievable! In many standardized breeds it is common for the standard to specifically penalize certain variants that may well be produced in the breed from time to time. This is especially true of color traits (red in Angus cattle, body spots in many horse breeds). Such exclusion can also be for physical traits such as a split scrotum in many goat and sheep breeds, or polledness or horns in many breeds of several species. Other breeds fault other traits considered by the association to be defects, such as su-

Breed standards should help breeders to visualize what is typical of the breed.

White Park cattle are generally white with black points, but white with red points, solid black, and solid red persist in small numbers and are a part of the heritage of this rare British breed. Photo by D. P. Sponenberg.

pernumerary teats in goats or an overshot or undershot jaw in several species.

Restrictive breed standards can have the unintended, and potentially dangerous, side effect of narrowing genetic variation because certain variants are excluded from the breed. For numerous breeds with great genetic breadth this narrowing is probably of little or no significance. For rare breeds whose population sizes are barely viable the standard must be a part of breed conservation strategy, which focuses on the larger issue of breed survival and less on the presence of incidental variation. This does not hold for fitness traits (lethal or debilitating conditions, for example) but certainly does hold for more trivial cosmetic traits such as color.

Welsh cattle lost many color variants as they became standardized, but this breed fortunately still occurs in numbers sufficient to form a viable population. White Park cattle, already in very low numbers, could easily slip further as a result of draconian measures against solid-colored animals. Randall cattle, nearly all of which are blue roan linebacks, have fortunately embraced and encouraged the use of the few black, red roan, and red individuals in the breed. Other breeds, such as the Navajo-Churro sheep breed, have embraced and celebrated the variation present in the breed, including rare variants such as polled rams. The result is that bloodlines of potential future benefit to the breed have been conserved.

In most breed standards some traits or variants are considered faults or may even disqualify animals from inclusion in the breed. The philosophy behind this determination varies breed to breed. If too lax, the result can be unsound or poorly conformed animals. If too strict, the play of variation in the population can be too restricted to allow for effective breeding strategies focused on long-term survival of the breed.

With the advent of genetic probes to test for certain traits, it is now possible for breeders to eliminate breeding animals that carry off-standard traits. Chestnut coat color, for example, is not allowed in registered Exmoor ponies, Cleveland Bay or Friesian horses. In the past this was of little consequence, as carriers of chestnut are rare in these breeds and they cannot be detected visually. With the advent of genetic testing, however, carriers can be detected and the temptation is to act against carriers as well as those horses that overtly express the gene in their coat color.

Culling an animal from reproduction removes its entire genome, not only an individual gene.

In that event the already small populations of these breeds will be further diminished by the culling of animals from reproduction on the basis of this single gene. It is important to remember that when an animal carrying chestnut, or any other single gene characteristic, is removed from the breeding population it is not only the single gene that is removed – it is the entire animal with all of its genetic makeup. For many rare breeds it is wiser to be less restrictive in culling on traits unrelated to soundness in order to assure breed survival.

Similar tactics of genetic selection based on single genes have been used in sheep breeding for control of the degenerative neurologic disease, scrapie. This disease has a profound genetic link, and the approach of most European countries and some states within the USA, is to insist that all breeding rams eventually have resistant genotypes. In some breeds, such as longwools, this is relatively easy because most sheep already have resistant genotypes. In other breeds, including many highly adapted breeds, resistant genotypes are reasonably rare and the result of the genetically-based selection programs is to remove many rams (and many bloodlines) from the male line of several breeds.

To muddle the issue further, evidence is emerging that sheep of scrapie-resistant genotypes can harbor the scrapie agent but simply not show evidence of disease. This apparent absence of disease is actually due to prolonged incubation periods rather than true freedom from the scrapie agent. The agent can then express itself as disease when it connects with a susceptible animal. To assure freedom of the agent, some countries such as Australia and New Zealand (both of which are free of scrapie) favor the approach that all imported sheep be from susceptible genotypes without evidence of disease. By this mechanism they are assured of not bringing in the agent. In the United States the approach varies breed by breed, and state by state, but the regulatory environment could change and eventually be dictated by federal government edict. Further, there

Lard-type hogs were once common in production agriculture, but became rare due to consumer demand for lean meats and a decreased value of lard as a cooking fat.

are several strains of scrapie. If all sheep are selected to be of the same resistant genotype, all could be susceptible to another strain of scrapie. Hopefully all sides of this complex issue will be weighed before strict regulations are established. The key point is that selection programs based mostly on single genes can have devastating and unforeseen effects on a breed and its future viability.

Breeds can be lost in various ways, and breed standards can affect at least two of these. But the most common cause of breed loss is unrelated to standards. It is outright extinction through lack of purebred recruitment. Narragansett Pacer horses simply vanished because breeders no longer produced them, switching to other horse breeds instead.

A second avenue to breed loss is related to standards. This is the more insidious loss that can occur through such drastic change in type (usually by ignoring or modifying standards) so that the original genetic package is gone even though the name remains. Modern lean swine breeds are dramatically different from their exceedingly fat lard-producing ancestors. Even though they are known by the same breed name, the genetic package beneath that name is drastically changed from the original. While the practicality of the change can be debated, the fact remains that changing standards and selection pressures do indeed change breeds.

Breed standards should be very carefully constructed.

A third manner in which breeds can be lost is for small populations to be managed so stringently and strictly that eventually the genetic variation needed for viability is no longer available to the breed, and it succumbs slowly and inexorably to extinction through loss of usefulness, vigor, and viability. Examples of this are hard to cite, but some dog breeds such as the Dalmatian dog may be approaching this situation as most now have disorders of purine metabolism with no genetic remedy present within the purebred population.

Breed standards should be carefully constructed because they strongly affect breeds and their breeding. Deciding on what is desirable and what is undesirable is controversial in most breeds. It is important for breeders to be clear-headed in establishing what are the most important traits that are central to breed identity as well as to breed function. Traits that are minimally important to true breed character are best left out of standards.

Breed Type

Each breed has a unique phenotype, which is a combination of appearance, performance, and behavior. Some of the most superficial elements are the easiest to quantify and describe, such as color, size, general shape and conformational peculiarities. All of these elements go together to yield a "breed type" which is the overall appearance of the breed.

"Type" is very difficult to define, but includes all aspects that make breeds unique. "Type traits" are those characteristics that set one breed apart from others in the same species. The individual animals of any breed vary in "typiness," which is the relative degree to which individuals express the type traits, and therefore represent the breed in its uniqueness within its species. Animals that strongly express the breed type are generally referred to as "typey" and are difficult to misclassify into any breed but their own due to their expression of breed-specific characteristics. Typey animals also have a subtle but important overall appearance that stamps them as not simply randomly bred.

Breed type is what sets one breed apart from others.

Breed type is critically important to the conservation of a breed, and is a concept that is increasingly ignored in modern animal breeding and evaluation. The trend in most classes of production livestock is to encourage most or all breeds to converge onto a single generalized type. This trend is largely being driven by the industrialization of livestock and poultry breeding, and is a response to market pressures for large quantities of a narrowly consistent product. Warmblood horses, beef cattle, swine, and dairy goats are all good examples of classes of livestock where the type has tended to converge across breeds. The result has been that many animals of several breeds are nearly identical for type,

Gyr cattle have a very strong breed type, making them difficult to confuse with other breeds. Photo by D. P. Sponenberg.

Nubian goat type signals an underlying genetic package that is predictable. Photo by D. P. Sponenberg.

and the underlying uniqueness of the various breeds has been compromised. Most of this has been accomplished by selection, rather than by crossbreeding – but crossbreeding has certainly been used in some instances to erase breed type. Warmblood horses, beef cattle, and dairy goats are all examples of classes of livestock where individual breeds are losing (or have already lost) their distinctive type because breeders have selected animals to converge upon a single generic type.

An odd quirk of breed type for many species of livestock is that type centers on the head, which is generally the only part of the animal that is not directly consumed or shorn. Heads do, obviously, have importance in being the major thinking and food-gathering organs of the body, but have even further importance in overall breed type. Head character, shape, and carriage are important signals of breed type in most species and are often the most breed-specific component of an animal. Horns and ears can be especially important in contributing to breed type, and while they are not directly involved in production they are important clues to the genetic background and breed affiliation of animals. The shape and set of a Nubian goat's ears, for example, can be confused with few other breeds.

"Type traits" set one breed as distinct from all others. Breed standards should emphasize these traits above all others because they most reflect the breed as a genetic resource. Evaluation of type traits presents many observers with a philosophic problem because type traits are usually not production traits. However, the presence of type traits indicates clearly that animals are of a given breed, and through that they also serve to indirectly predict production characteristics that go along with that breed. So, although type traits are usually not direct production traits, they are indeed indicators of production by virtue of characterizing breed identity.

Among goats, for example, Nubian goats betray their breed identity by their conformation and type traits (roman nose, long ears, dairy conformation). A buyer can expect a goat with good fertility, reasonably nonseasonal breeding, and production of ample quantities of rich, sweet tasting milk from this goat – which is also likely to be minimally aggressive to other goats, and to vocalize a lot. All of this information is part of breed identity, and the breed type has betrayed it to the observer, even though most of these traits of production interest have been delivered by the underlying genetic package that cannot be observed directly. A Texas Longhorn cow, in contrast, betrays her breed identity by horns, ears, body, and head conformation. That package of characteristics should ideally go with a cow likely to rebreed quickly after calving, raise calves well, forage well, resist droughts well, and do all of this over a very long life.

General conformation, production, and soundness traits are more likely than type traits to be shared across several breeds. These sorts of traits are essential to animal function and well-being, and as a result these receive considerable attention in breed standards. The boundary between these traits and type traits can be blurry, but certainly they do overlap.

Realizing the difference between type traits and soundness and production traits as they relate to breed character can be essential in understanding how the standard relates to the maintenance of a breed. Soundness and production traits usually answer the question of "how good is this animal?" while type traits answer the question of "how typical a representative of this breed is this animal?" Those two questions are different, and confusing the two muddles breed distinctiveness and management. Both questions are essential when evaluating animals, but each is targeting different information.

Type traits and production traits are usually different.

Although type traits are the most essential for breed character, it is also true that conformational and soundness traits are important. In both sorts of traits it is important to allow some leeway for variation, but at the same time animals deviating greatly from the norm have little or no role in breed conservation or breed maintenance.

Qualitative and Quantitative Traits

Breed standards usually dictate both the qualitative and quantitative elements of breed character and type. Qualitative elements are "either/or" traits such as coat color and length, class of wool, presence or absence of horns, and similar characteristics that are usually easy to classify as to their presence or absence in an animal. Most standardized breeds have a limited array of choices in any of several qualitative traits. Included among these are a tendency for most standardized breeds to limit the color array (frequently to a single color or pattern), and a single choice for horn presence or absence. Variation in qualitative traits is usually higher for landrace populations than it is in standardized breeds. This subtlety often results in the uninformed observer failing to recognize landraces as legitimate breeds, for it is difficult to see past the superficial variation in color and horns to detect the underlying conformational and performance similarity.

Qualitative traits are "either/or" traits. Quantitative traits are "how much" traits.

Quantitative traits are those that vary along a continuum, such as height, weight, quantity of milk production, linear measurement traits, and similar "how much" traits. Quantitative traits are as important to breeds as qualitative traits, but are much more difficult to capture in a breed standard. Certainly size parameters can be described, but traits such as growth rate, milk production, fecundity, and egg size are equally critical to defining a breed and yet are much more difficult to detect by casual one-time observation such as occurs in a show ring.

Traits such as hardiness, parasite resistance, fertility, and other complex characteristics are even more difficult to classify as either quantitative or qualitative. These traits frequently have contributions from a wide variety of genetic mechanisms, some simple and some more complex. These are among the hardest of all traits to describe or evaluate, but are also among the most important for many breeds of livestock. These traits of adaptation and function can be especially important in landraces.

Changes to the Breed Standard

Breed standards should be thoughtfully and carefully constructed from the outset, so that changes should be rarely needed or advised. Experience and improved information eventually might well indicate that some changes are needed. Changes should be agreed upon by general membership in an association, and are best accomplished following an extensive educational campaign concerning the breed, its history, its type, and the potential effects of the changes on the breed as a genetic resource.

Breed standards should be realistic portrayals of breeds.

It is especially necessary for breed standards to reflect the biological realities of the breed. Breed standards should be realistic portrayals of the breed's appearance and optimal biological function. The ideal, or standard, should reflect the reality of the breed so that pure breeding populations can be achieved that meet the standard.

Changes in breed standard are always highly charged politically, because these changes necessarily indicate a change in the direction that the breed is being taken. This also means that some breeders come out ahead because they own animals more in keeping with the revised standard, while others are disadvantaged by the change because they own animals less in keeping with the revised standard. The political and economic aspects of such changes can become very highly contentious, and need to be handled openly and fairly in order for the breed and its breeders to not suffer.

Breed Type Reproduces Breed Type

For breeds to be useful genetic packages, they must be genetically consistent to predictably reproduce themselves. Breeds breed true. The standard and the rules and regulations of the association should all be targeted toward the goal of assuring that the breed maintains its status as a consistent genetic package. Standards serve their breeds very poorly if they fail to define breed type, overall performance, or conformation adequately. Such loose standards result in the breed becoming too loosely defined to serve as a genetic resource.

Breeds breed true.

Breeds must also be subjected to ongoing selection that reflects and emphasizes breed type. For any breed to maintain a useful status as a genetic resource it must maintain its genetic integrity. While genetic integrity can be lost through several means, selection away from breed type is one common avenue to lost breed integrity.

Breed type should always reproduce breed type, leading to readily identifiable groups of animals such as these Pomeranian geese. Photo by D. P. Sponenberg.

4. Maintaining Breeds

Maintaining breeds as healthy and viable genetic resources should be the goal of each breeder and each breed association. This effort can and should involve different breeding strategies, both within breeds and across breeds. Each breeding strategy has different consequences for the individual herd as well as for the breed. Each breeder must tailor a strategy for the specific mix of philosophies, situations, and goals that are unique to the herd he or she is breeding. No single strategy fits all situations, but each strategy is a wise choice for certain goals and production systems. No "single best answer" exists for breed and herd maintenance. Having a variety of approaches, all within a single breed that is bred pure, works well for breed conservation and maintenance.

Understanding the different breeding strategies (inbreeding, linebreeding, linecrossing, outcrossing, and crossbreeding) is very important for conservation breeders. These terms all have slightly different definitions to different groups of breeders, but the key fact is that the pairing of animals for reproduction has varying outcomes depending on the relationship of the animals mated. The results of each of the different breeding strategies are subjective points along a single line that vary from completely inbred and uniform at one end to completely crossbred and variable at the other end. Just exactly where along this line to draw the boundaries between inbreeding and linebreeding as well as linecrossing and crossbreeding are subjective, but the effects of these strategies on populations are very real. Each strategy has an important role in shaping animal populations.

Maintaining breeds as genetic resources should be the goal of each breeder and each association.

Breeding Strategies

Linebreeding and Inbreeding

The breeding strategies of linebreeding and inbreeding are only different from one another in degree. Both of these involve the mating of related animals. The related animals can be either distantly related or closely related – any mating of related animals is technically inbreeding. As a more practical definition, in-

Randall cattle have had a history of intense linebreeding. Photo by D. P. Sponenberg.

breeding can be arbitrarily set as the mating of first-degree relatives. First-degree relatives include offspring, parents, and siblings. Although this is only one possible definition among many, it usefully separates inbreeding from linebreeding at a specific level. Linebreeding can then be considered as the mating of related animals, but of less close relationship than first degree. Matings of aunt to nephew, grandparent to grandoffspring could all be included here, as well as more distant matings such as cousins.

Linebreeding and inbreeding differ from one another only by degree.

Linebreeding and inbreeding both result in increased genetic uniformity of offspring, especially if adopted as a long-term strategy with appropriate selection. Linebreeding increases genetic uniformity because parents are related and therefore descend from a common and limited gene pool, which means that they have limited genetic variation to pass along to their offspring. Uniformity of appearance and performance of linebred animals springs directly from this fact. The uniformity can be for very good looks and performance, or for very bad looks and performance – the starting animals as well as selection practices determine the relative quality of the end product. In addition, the degree of relationship of the parents helps to influence the degree of uniformity in the offspring. The closer the relationship the more uniform the offspring. And, over several generations of inbreeding, relationships become ever more close, and uniformity also increases.

A very important historic note is that linebreeding and inbreeding were and can still be usual strategies for the establishment of breeds. These two breeding strategies increase uniformity, and therefore predictability, of any population of animals. The very essence of a breed is sufficient uniformity for predictability. In that regard, both inbreeding and linebreeding can be effective in achieving the status of animal populations as true genetic breeds.

Inbreeding and linebreeding increase uniformity and predictability.

The strength of linebreeding is that it increases homogeneity and predictability. When coupled with selection (which it usually is) the result is a productive, predictable gene pool. This is the key importance of a purebred animal – predictability of production. Potential problems can occur in linebred populations and are even more common in inbred groups. Common problems include loss of general vigor and especially loss of reproductive performance. In addition, inbred groups can have a concentration and expression of undesirable recessive traits. Skilled selection can help offset these so that several linebred and inbred resources (that is, breeds or strains within breeds) are indeed productive, vigorous, and reproductively sound. Randall cattle are an excellent example of a viable, though inbred, resource. However, due to the potential negative aspects of inbreeding, it is important to manage all inbreeding within limits. It is also wise to assure that outcrosses to other lines within the population

Every animal in every breed should have a potential outcross mate.

are available in the event that production and reproduction begin to suffer in any linebred group. Every animal in every breed should have a potential outcross mating available, or else the breed will find itself perched on a dangerously narrow genetic foundation.

An extreme example of inbreeding was a Texas Longhorn cow from a line that was resistant to inbreeding depression. This cow had the

This linebred Lely strain Texas Longhorn cow is 7/8 (87.5%) the influence of one bull which was her sire, grandsire, and great grandsire. Photo by D. P. Sponenberg.

same bull as sire, grandsire, and great grandsire – so she was 87.5% the genetic influence of that one animal. She kept producing calves at intervals of a little over 10 months during her long and useful life. But she is the exception to the general rule that close inbreeding results in depression. Usually this sort of animal results from breeders not planning for future generations, and only rarely is such a highly inbred animal productive and adapted.

Inbreeding is not without risk, and needs to be acknowledged and planned for in any long-term breed maintenance program. Both linebreeding and inbreeding expose and concentrate recessive traits, whether these are positive or negative. These two strategies must only be employed with a firm and resolute commitment that deleterious traits be rigorously culled out of the breeding population.

Outcrossing: Crossbreeding and Linecrossing

Outcrossing is a philosophical and biological opposite of linebreeding, and involves the mating of animals that are not related. Outcrossing and outbreeding are synonyms, and the results are commonly characterized as outbred (as contrasted to inbred). Outcrossing can be subdivided into two subgroups: crossbreeding and linecrossing. Crossbreeding is the mating of animals from two different breeds. Linecrossing, in contrast, is the mating of unrelated animals from within the same breed. Usually these matings occur between animals from two different lines within the breed, which leads to the term "linecrossing."

Crossbreeding is a fascinating phenomenon, partly because different results occur depending upon which stage of crossbreeding is considered. The first stage is the initial cross. A useful example comes from cattle. When Angus and Hereford cattle are crossed the initial result is a uniform crop of "black baldy" calves. The calves exhibit the dominant traits of both breeds: black and polled from Angus, and white face from Hereford. These offspring have benefited from the specific combination of the genetic array of the parental breeds, and each calf gets half from each breed, and therefore every calf is pretty much like the next. This generation is called the F1 (for first filial) generation. Each parental breed is uniform. Each calf gets half of its genetic makeup from each parental breed. This first calf crop is reaping the benefits of homogeneous parental

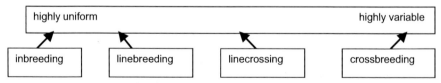

Relative degrees of genetic relatedness in different breeding strategies.

"Black baldy" cattle are typical crossbreds – great productivity but little predictability in their own reproduction. Photo by Terry Swecker.

breeds, as well as hybrid vigor resulting from the mating of two unrelated parents.

If these crossbred calves are in turn mated to each other (resulting in the F2 or second filial generation) variability then increases because the calves of the next generation are varying mixes of the half one thing, half another thing represented in their parents. The color, let alone other characteristics, can serve to illustrate what happens when a crossbred animal is used for reproduction. Interbreeding black baldies produces black, black baldy, red, and white-faced red calves. The initial consistency present in the F1 generation is gone, and the result is a variable group of calves.

This variability is not all bad. When combined with selection the excellent individuals can be skimmed off the herd and used to good advantage in the show ring and other situations. They may indeed have excellent type and performance. What they lack, though, is the ability to consistently pass along this excellence to the generations following that initial F1 generation. They are something of a

Body Color	Head Color	Horns
black	white	polled
black	white	horned
black	black	polled
black	black	horned
red	white	polled
red	white	horned
red	red	polled
red	red	horned

Variations in calves produced from black baldy to black baldy matings.

dead end for a breeder, even though in themselves as individuals they may be wonderfully productive animals.

Many of the advantages arising from a crossbreeding strategy (increased vigor and reproductive efficiency) are diminished under a linebreeding strategy. Conversely, the advantages of using a linebreeding strategy (consistency and predictability) are diminished under a crossbreeding strategy.

Crossbreeding is the mating of animals of two different breeds. Linecrossing is the mating of unrelated animals from the same breed.

Linecrossing is less extreme than crossbreeding, because it occurs within a single breed. While it has some of the same biological consequences as crossbreeding, the crossing is contained within a single breed, providing the benefits of crossbreeding without loss of breed character and type. As a result, the variability is not as great as a cross between breeds, so the boost from hybrid vigor is not as great. This technique can be used to good advantage in certain breeds, such as the Angora goat, where interbreed crossing would make no sense at all because no other breed could be crossed into the Angora without causing the loss of its distinctive and valuable fleece characteristics. Linecrossing can contribute to the vigor associated with high production, but lowers the consistency of production by virtue of creating animals that are more genetically mixed than linebred animals. Only a careful analysis of each individual situation will indicate whether or not this is a good trade-off.

Angora goats are one breed where crossbreeding has little to offer, but linecrosses can yield some of the same benefits. Photo by D. P. Sponenberg.

Defining Matings as "Related" or "Unrelated"

The exact point at which matings are classified as "related" and contributing to inbreeding, versus "unrelated" and not contributing to inbreeding is impossible to define in absolute terms. Nearly all purebred matings are more closely related than would be matings to another breed. This, after all, gets back to the whole point of breeds being genetic resources that are genetically uniform enough to be predictable.

Animals that have no ancestors in common back to grandparents can be considered as unrelated for outbreeding purposes.

Animals are unrelated if none of the ancestors on the sire's side also occur on the dam's side. As a useful general rule, the mating of any two animals that have no ancestors in common back to grandparents can be considered as an outbreeding, so that the offspring has no ancestors in common on sire's side and dam's side back to the greatgrandparents. That is, any relationship at this level is generally trivial and contributes so little to an inbreeding coefficient that it is generally safe to ignore. At the other extreme, the mating of

Pedigrees of a male and several female mates demonstrating varying levels of inbreeding and linebreeding. The ancestors that the females have in common with the male are in italics.

male A	sire B	sire C
		dam D
	dam E	sire F
		dam I

mating to female E is inbreeding (son to mother)

female E	sire F	sire G
		dam H
	dam I	sire J
		dam K

mating to female L is inbreeding (full brother to full sister)

female L	*sire B*	*sire C*
		dam D
	dam E	*sire F*
		dam I

mating to female M is more distant inbreeding (half siblings)

female M	*sire B*	*sire C*
		dam D
	dam N	sire O
		dam P

mating to female Q is linebreeding (cousins)

female Q	sire R	*sire C*
		dam D
	dam N	sire O
		dam P

mating to female S is linebreeding (nephew to aunt)

female S	*sire C*	
	dam D	

mating to female T is linebreeding (grandsire to granddaughter)

female T	sire U	*sire A*
		dam V
	dam N	sire O
		dam P

mating to female W is an outbreeding (no relationship of mates)

female W	sire X	sire Y
		dam Z
	dam N	sire O
		dam P

first-degree relatives (parent to offspring, full and half siblings) can be considered inbreeding because these animals are so closely related. Between these two extremes lies linebreeding, which varies in degree but still lies at a level below severe inbreeding and above outbreeding.

Linebreeding or Outcrossing: Which Is Best?

Different breeders have different philosophies and goals. Philosophies might include strict conservation principles, strict commercial utility, high-profile program with name recognition, or several others. Goals might include excellent temperament, maximum weight gains, production on minimal inputs, easy births, prolificacy, and many others. As a result of different philosophies and goals (which can all be legitimate), the phenomena associated with crossbreeding and linebreeding have different consequences for different breeders because these strategies each work well for different goals. The specific goals and philosophies are important in driving decisions as to which strategy to use, and these legitimately vary among breeders of nearly all breeds. Underlying philosophic questions (why is the breeder breeding livestock in the first place?) are essential for all breeders, but are frequently not asked. In the absence of a guiding philosophy and set goals, breeding programs generally fail to make much progress toward any goal. All breeders should develop a guiding philosophy, for this assures better progress in a more focused breeding program.

Different breeders have different goals.

Inbreeding generally firmly establishes traits in the offspring. Inbred offspring, ideally, are more consistent in reproducing their own type than are outbred individuals. Breeders can use this to good advantage. However, breeders must keep in mind that prolonged multigenerational inbreeding will likely bring a decline in vitality and reproductive fitness. The real risks of long-term inbreeding in no way limits the usefulness of short-term, targeted inbreeding to accomplish specific goals within a herd or breed. Specific strengths and uses for inbred animals are in outcrossing to other lines in an effort to balance genetic founders within a breed or herd, increase vigor and vitality, or to reap the benefits of some specific trait in a given line. For example, the Randall cattle breed traces back to a very few founders. Using bulls that are linebred to only one of these makes it possible to assure calves that are relatively high percentage breeding of that founder so that the founder's influence is not lost to the breed. In contrast, by using non-linebred matings it is impossible to avoid diluting the founder contribution to the point that no individual cattle of the breed have any high percentage of any founder, and all cattle become related to one another.

Uniformity of progeny is important to commercially based breeding operations. In a purebred setting linebreeding is one strategy to achieve this end. Reasonably uniform animals that perform predictably are of great value to farmers, enabling them to target management for the expected level of production that the animals are going to achieve. Obviously the animals are never going to be entirely uniform, and the better producing animals will always be retained in favor of the lower producing ones. However, as the variation diminishes, the top and the bottom performers of the population approach one another (hopefully by the bottom coming up toward the top), so that the casual viewer is struck by how much the animals resemble one another and make a uniform group.

Linebreeding takes time and commitment, while linecrossing and crossbreeding can be quick fixes and are tempting strategies for a variety of reasons. One outcome of crossbreeding is initial phenomenal results, especially if the parents that are recruited for the crossing are thoughtfully selected for complementary strengths. The boost of crossbreeding comes from hybrid vigor, and can easily be seen in several breeds of many species – especially those used for meat production. The boost in overall vigor is used to very good benefit in several production systems, and some examples illustrating this are discussed below.

Crossbreeding does not make sense if the goal is consistent production generation to generation. Crossbreeding does indeed make sense in many other circumstances. One of these is the production of a terminal animal. Show ring

Crossbred cows, such as this Galloway x White Shorthorn, are useful in commercial settings due to the boost they get from hybrid vigor. Photo © Evelyn Simak.

animals and meat producing animals provide useful examples of situations in which the boost from hybrid vigor can be put to very good use. One very well established system is the British system of mating hill breed ewes (small, adapted, with less growth rate) to Longwool rams (large bodied, heavy fleeces, good maternal characteristics and usually prolific) to produce crossbred vigorous ewes for production in more favorable environments. These crossbred ewes are medium to large, fertile, prolific, and shear adequate amounts of good wool, having inherited a combination of production and adaptation traits from their divergent parent breeds. The crossbred ewes are mated to large, fast growing meat breeds to produce large numbers of fast growing lambs for the market. The lambs of this last stage are not retained for breeding, because at the end of this three-breed cross the genetic variability is such that they would not reproduce consistently enough to be useful. A similar system is used with Galloway cows and White Shorthorn bulls to produce blue roan crossbred brood cows that cross well with terminal beef sires.

Successful crossbreeding depends on the continuous availability of purebred stock.

Note well, though, that crossbreeding systems all depend on a reliable, long-term continuous source of the female line as well as the male line. Crossbreeding requires different and divergent pure breeds for its success. Without pure breeds, crossbreeding quickly fails. The Cleveland Bay horse is falling victim to the success of its crossbred offspring. Cleveland Bay mares are frequently crossed with Thoroughbred stallions to produce valuable sport horses. The purebred Cleveland Bay foals have less market value than the crossbreds, so that purebred recruitment is low and could eventually lead to the point where the cross is no longer possible due to loss of the purebred Cleveland Bay mares.

Industrial egg laying chickens are produced in a very clever scheme that manages genetic material for maximum production as well as maximum corporate security over genetic resources. The primary breeding lines are very highly selected, and very closely guarded. These include two that are crossed to make up the male side of multiplier flocks, and two that make up the female side of multiplier flocks. The multiplier flocks are much less closely guarded than the primary breeders, because they are hybrids of two lines and therefore cannot be used effectively to regenerate the original resource. The resulting two groups of hybrids (one the source of males, the other the source of females) do cross well with one another to produce the final egg-laying hens. These benefit from great vigor and ability from their hybrid, outcrossed genetic makeup. They are superb layers of eggs of consistently high quality, but would be very substandard as breeding animals because their offspring would be too variable. As a consequence, the industrial breeders need to have no control over where the production birds go – the corporate resource is safely back at home in the grandparental lines. Pure breeds of livestock are a similar resource, to be guarded and not squandered on crossbreeding without replacing the purebred resource.

Crossbreeding basically "uses up" genetic material without contributing to its maintenance in predictable and useful breed packages. It therefore has little role in breed conservation despite its very useful role in animal production systems. The production value derived from crossbreeding ultimately depends upon those segments of agriculture that are committed to purebreds – more specifically, purebreds of different breeds for different production and market niches. The importance of this cannot be overstated, and all of production agriculture owes a huge debt to dedicated purebred breeders who maintain pure breeds as genetically distinct and useful entities.

Pure breeds of livestock are a resource to be guarded, not squandered by crossbreeding without purebred replacement.

Is crossbreeding bad? Absolutely not, although if breeders go for the fad of the moment they can use up adapted breeds by crossbreeding without purebred replacement. This leads to diminished choices for future breeders. An example of this is the feral goats of Britain that have been incorporated into cashmere producing systems. Some of these ferals may have been the remnants of the old type of North Atlantic goats, mainly of Scandinavian origin and with exquisite environmental adaptation. It is too early to say if something potentially useful has been lost, but it is not too early to say that we have indeed lost something. Similarly, throughout most of the Americas the adapted and productive Spanish-based Criollo cattle have now all nearly disappeared as a result of being crossbred out of existence. Included in this loss were several strains of Pineywoods and

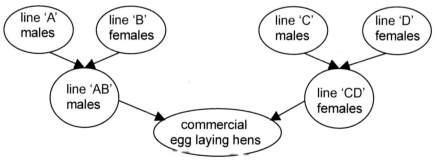

Genetic production of industrial egg producing hens

Florida Cracker cattle from the southeast United States. The crosses were generally to zebu-type cattle, and the first generation had phenomenally good production and vigor. Subsequent generations, though, began to lag in production until eventually they were generally below the level of the original Criollo base from which the crossbreeding began. At that point the genetic resource was gone, and is now impossible to recover. Crossbreeding is a very real threat to local, purebred, long-adapted landrace animal populations.

Pure breeding of different breeds each adapted to different production and market niches is essential to long-term agricultural productivity and survival.

Many of the older, high-reputation breeders of most breeds have indeed used a linebreeding strategy (coupled with selection) to produce the animals and breeds that breeders today highly desire. For most breeds, too few breeders have long-term commitments to linebreeding and the development of consistent, productive lines that are predictable for performance. A new generation of committed, knowledgeable breeders is desperately needed.

The benefits of linebreeding and linecrossing can both be exploited by alternating these strategies from generation to generation within a single herd. This is accomplished by assuring that linecross animals, rather than being further linecrossed, are preferentially linebred back to one of the specific lines that is in their own ancestry. Mating linecrossed animals back to one of their parental lines can do this. That sort of mating then produces linebred animals once again, which can be used in linecrossing to avoid multi-generation linebreeding with its attendant risks. While the resulting population will not all be intensely

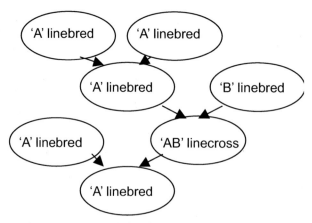

Alternating linecrossing with linebreeding generation to generation.

linebred, some portions of the overall population will indeed be moderately linebred. This portion will have the benefits of linebreeding, but without much risk of the drawbacks that can possibly plague more long-term, multigenerational linebreeding programs. A breeding program that accomplishes the alternation of these strategies needs to be carefully planned and executed, and the results will usually reap great benefits for both the breeder and the breed.

A breed population is best served if several breeders are using slightly different breeding strategies, philosophies, and methods. Breed genetic health benefits if some breeders are linebreeding and others are linecrossing. This allows for successful genetic combinations to be developed in a variety of locations and conditions. This is good for breeds and breeders. A single program and philosophy will not fit all situations, and breeders need to encourage diversity of approaches and techniques. This requires coordination and cooperation among purebred breeders, and overseeing and facilitating this is an important role for breed associations.

Breeds are well served by different breeders using different breeding strategies.

Linebreeding and Linecrossing as Strategies for Population Management

Managing small populations, including individual herds of any breed or entire rare breeds, is basically the management of inbreeding. One strategy for long-term management of inbreeding is to manage it carefully while assuring that outcrosses are available for every animal within the population. In that way it is possible to avoid inevitable inbreeding (and the likely depression it brings) by outcrossing occasionally. One strategy for this is outlined in the *Conservation Breeding Handbook* by Sponenberg and Christman, and available as a resource from ALBC. A summary of the approach is given here.

While the details can be variable, a minimal approach is outlined here. A first step is to subdivide the population into three groups. This is usually done by bloodline or family affiliation, but other strategies could be used. The rationale for subdividing the population is best if it reflects both numbers (to assure relatively equal numbers in each group) as well as genetic relationships (to try to assure maximal genetic distance between the groups). The groups are labeled "A," "B," and "C."

The specifics of the program demand that animals be labeled with their bloodline identity. To do this, any animal that is over 75% one line should be designated as that line, regardless of the remaining 25% of the breeding. Similarly, any animal with at least 25% the breeding of a foundation line should also have that included in its designation. So, an animal that is 50% A, 25% B, and 25% C could be labeled "ABC," while one that is 50% A, 44% B, and 6% C is simply "AB," and one that is 87% A and 13% B is simply "A."

The specific strategy for illustrating the principles of this system is to always use linebred sires. These are alternated in the herd, likely year to year, and both linebred and crossbred replacements are saved. If the dams are both linebred and linecrossed, and the sires all linebred, then the system functions to provide for both linebreeding and linecrossing in every year.

Management of a small population by managing three bloodlines.

Sire line	Progeny Produced from Each Dam Line						
	A	AB	AC	B	BC	C	ABC
Year One: A	A	A	A	AB	ABC, AB, or AC	AC	A, AB, or AC
Year Two: B	AB	B	ABC, AB, or BC	B	B	BC	B, AB, or BC
Year Three: C	AC	ABC, AC, or BC	C	BC	C	C	C, AC, or BC

The important detail in this management scheme is that the basic rules for line assignment allow the genetic material to move from linebred to linecrossed individuals, and then back into linebred individuals. So, for example, a linebred "A" dam mated to a linebred "B" sire produces a linecrossed "AB" offspring. This "AB" offspring can, though linecrossed, contribute linebred offspring to both "A" line and "B" line following mating to sires from those lines. Or, it could further contribute to linecrossed individuals by mating to a "C" line sire.

Rotation of sires through the entire herd assures that linecrossed and linebred animals are being produced every year. It is also important for breeders to assure retention in the herd of both linebred and linecrossed replacements every year. For example, failing to retain "AC" animals in year three could result in this category being lost (if temporarily) so that one portion of the distribution of the genetic material is missing.

This example of the conservation breeding program is the bare minimum, and larger populations can easily vary the basic program by having more lines or more breeding groups. Likewise, groups of breeders could band together and exchange males every few years (every two works well) to assure even less risk of inbreeding depression but without foregoing all of the benefits of linebreeding.

Important Applications of Linebreeding

Linebreeding has a few specific powerful benefits for many rare breed conservation programs. In rare breeds it is common to find that some bloodlines of a breed have dwindled down to a few individuals, and these are most often females. This is especially likely to happen in breeds that are widely used for crossbreeding, such as Florida Cracker and Pineywoods cattle.

A major strength of inbreeding is the fixation of traits in a given line of animals. Even extreme inbreeding can be used as a short-term strategy to accomplish the specific goal of making certain traits more prevalent in a breed or herd. An excellent older female, for example, with proven production, can be mated back to a son for an inbred replacement. If the offspring is a son, then he can be widely used throughout the breed to spread the superior genetic traits of the original female more effectively than would be possible without resorting to the close inbreeding.

This strategy can be especially effective when used to correct the underrepresentation of the genetic influence of certain founders of rare breeds. Some of these may be female remnants of important bloodlines that have declined as in the preceding example. Another common situation for using this technique is to enhance imported bloodlines that may have only come through a single sex of animal – more commonly a male imported via semen, but occasionally a female of a rare bloodline.

In situations where a line has dwindled to a few females (and as a practical issue these females are generally closely related) one good strategy is to mate the

This productive Tennessee Myotonic doe survived to great old age, and was mated to sons to produce linebred sons that could be used to disperse her genetic strength more widely. Photo by D. P. Sponenberg.

females to a purebred male from another line (this is by definition a linecross), then mate a male offspring back to the original females. The result is a crop of youngsters that is 3/4 the original line. These "3/4" males can be used back on the original females, and this strategy (young males, older females) can be used until the older females cease being productive. The goal is to take genetic material that can only be used to a limited degree by virtue of being in female form, and generate males that can be used more widely to distribute the genetic material in these rare foundation bloodlines broadly throughout other portions of the breed. It is important to note that while these initial linecross (first generation) and linebred (succeeding generations) males are being produced, so are females that can be effectively used in other portions of a breeding program without endangering the critical role that the foundation females are playing.

Similar strategies can be used with outstanding females of any breed, but especially for rare breeds. Outstanding, or genetically unique, females can be mated back to sons to produce offspring (hopefully male) that are 3/4 the influence of the original outstanding female. This has been done in many rare breed conservation programs with individual rare line females. While this strategy cannot be used over several generations without running into inbreeding depression, it is usually a successful strategy for a generation or two, at which point the males produced can be linecrossed to other lines and thereby contribute widely to the breed.

For example, within the Colonial Williamsburg Foundation importation of Leicester Longwool sheep came a single ewe from the Glendhu property in Tasmania. She was unrelated to all the other imported sheep. This ewe was conformationally correct for the breed, strong-woolled (as appropriate to the breed), highly maternal, and reproductively fit. On two occasions she was mated back to different sons, and the results included two rams (as well as ewes) that were

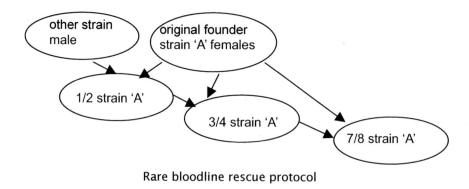

Rare bloodline rescue protocol

This is the Glendhu ewe whose linebred sons proved useful to the Leicester Longwool conservation breeding effort in the USA. Photo by D. P. Sponenberg.

3/4 her genetic influence. These rams were then moved to other flocks so that it was possible to more widely disseminate the founder ewe's excellence in type and production throughout the breed in the USA.

This ewe's usefulness to the breed was only possible because of the past efforts of long-term breeders of purebred livestock who make such linebred animals available. Every breed must have such breeders and such animals. In general the line becomes associated with the breeder, such as in the case of the Ridley Bronze turkey, the Conway Pineywoods cattle, and Hillis Cotswold sheep. It is the long-term dedication of individual breeders that provides these distinctive genetic packages that are so useful for others to build upon. Each breed and each generation within the breed needs breeders with this dedication in order to safeguard the genetic heritage of the breed.

Inbreeding and Loss of Diversity

Inbreeding has several important consequences. One is inbreeding depression, which refers to the decline in vigor of inbred animals as compared to outbred animals. Inbreeding usually diminishes overall vigor as well as reproductive success. This occurs at variable rates in different populations so that the significance of this phenomenon as a practical issue varies and exceptions to the general rule do occur.

Inbreeding depression is a loss of general vigor and reproductive success.

A second and important consequence of inbreeding is that it can cause small populations to lose genetic variation. This can be good for predictability. As animals become more genetically uniform they also tend to become more similar in looks and performance. Predictability is the hallmark of pure breeds, and so this can be a good consequence of reduction of genetic variability. The down side, though, is that genetic variability is essential not only for population health, but also for providing the raw material for selection and improvement. Highly inbred populations may lack sufficient variability for selection efforts to make any progress in production.

If all animals in a breed or herd are closely related, inbreeding and inbreeding depression become inevitable.

Even though some lines of some breeds withstand inbreeding very well, inbreeding depression is a widespread and well-documented phenomenon.

All breeders should manage their populations to assure that problems of inbreeding are avoided. Breeds should be managed so that every animal within the breed has an outcross available. This strategy assures that a "back door" escape is available to the breed should inbreeding depression become evident. Inbreeding and its consequences can then be quickly remedied by resorting to an outcross. The situation most to be avoided is the one in which all animals within a breed or herd are closely related to one another so that inbreeding becomes inevitable. This is a common trap for rare breeds when attempts are made to "change the ram" every year by cross-country trading. This strategy results in all flocks becoming related through having all used the same males, and it therefore becomes impossible to correct for inbreeding depression should it become manifest in the breed.

Genetic variability and genetic uniformity play tug of war. Populations at one extreme are so variable that they are completely unpredictable as to type and production. Populations at the other extreme are so uniform that vigor diminishes and genetic selection is impossible because all animals within the population are so similar to one another. Most breeds lie between these extremes and must be managed so that the benefits of predictability are not lost to diminished vigor on the one hand, and so that enhanced vigor is not gained at the expense of low predictability, on the other hand.

Breeds should be managed so that every animal has an outcross.

An example comes from Tennessee Myotonic goats. A 15-year-old, productive doe was mated back to a son to produce a buck that was 3/4 the genetic influence of the original productive doe. This inbred buck never grew as big as his herdmates, but he consistently produced exceptional kids when he was used for linecross matings. This example shows both inbreeding depression and hybrid vigor, and illustrates that inbred animals are likely to outproduce their own performance when used for linecrossing.

This inbred Tennessee Myotonic buck sired linecross kids that consistently outproduced his own record. Photo by D. P. Sponenberg.

Monitoring Inbreeding

Monitoring the generational increase in inbreeding is important for small, isolated populations such as rare breeds. Monitoring inbreeding is relatively easy if database registry software is used to track registrations and pedigrees. Any other approach is likely to be too time consuming to be practical for most registries and associations.

While not strictly a generation-to-generation evaluation, a more practical approach is to compute breed-wide inbreeding coefficients annually. These can determine a breed-wide trend. In addition, an overall range of inbreeding coefficients can also be computed.

More difficult to assess than an overall breed-wide average degree of inbreeding is the status of the degree of inbreeding within the various individuals of the breed taken one by one. If animals in one year are highly inbred but not all to the same individuals within the breed, then it is possible to lower the inbreeding coefficient in subsequent years by selectively mating unrelated animals. In contrast, if all of the animals are inbred to the same individuals, then such corrective measures are impossible. As a consequence of this, the specific character of the inbreeding is of equal concern as the overall degree of inbreeding. If the inbreeding is occurring in different directions (to different individuals) then the breed is less precariously perched than if the inbreeding is all occurring to the same few individuals across the entire breed. Both the degree of inbreeding across the breed as well as the degree in individual animals are important. To most effectively track this detail, it is necessary to evaluate the degree of inbreeding with respect to several different founders instead of using a single average across the entire breed.

Randall cattle were rescued from a very small foundation herd, and all of the founders were related. A breeding program was undertaken to specifically assure that cattle that were very high proportions of certain founders were mated together (linebreeding) in order to assure a high percentage of each founder in certain animals. Ideally these would be bulls, as semen could then be collected. These "high percentage founder" animals could then be mated across the breed, usually to animals of lower percentage of that specific founder, in order to assure a relatively even distribution of the few founders across the entire breed. By concentrating the proportion of the different founders in different cattle it was possible to assure that matings were available that were relatively unrelated, rather than having all of the cattle equally related across the entire breed.

Overall inbreeding and inbreeding relative to single founder animals are two different and important issues in rare breeds.

Analyses of the finer points of the degree and direction of inbreeding are more difficult to obtain than are overall degrees of inbreeding across the entire breed. Some databases will allow analyses to compute overall relatedness to specific individuals. These techniques are usually used when breeders are planning to purchase a new breeding animal, so that the new animal can be compared to the existing herd to determine the inbreeding consequences. These approaches can be used more broadly to determine the overall relatedness of a breed to a few key individual animals that may be swamping the breed genetically. Not only is the overall degree of relatedness to those individuals important, but also a more

Randall cattle remain productive and viable despite a long history of linebreeding. Photo by D. P. Sponenberg.

detailed individual-by-individual analysis to assess the range of relatedness should be considered. An example is that an overall degree of relatedness to an individual (usually male) in a population might be 25%. A range of relatedness across individuals within the breed of 20% to 75% is very different from a range of 0% to 50%. In the latter case, outcrosses are still available. In the first case, they are not available and the breed has hit a true bottleneck in that individual because all matings must now be linebred to that individual.

Monitoring Effective Population Size

Effective population size is a relative measure of the number of truly different genetic individuals in a population. For example, a group of 10 full siblings represents a lower number of genetic individuals than does a group of 10 unrelated animals because the genetic variation is much lower in the family group than in the nonfamily group.

Some breed associations place great value in estimates of effective population size for their breeds as a rough measure of its genetic health and the degree of genetic variation. Effective population size can be a useful estimate of future inevitable inbreeding trends, because low effective size indicates a future in which all animals will be related and therefore all matings will be inbred matings.

Effective population size is most simply expressed as: 1/effective population size = [1/(4 x number of males) + 1/(4 x number of females)]. The numbers of males and females are the numbers of reproducing males and females. The mathematics of this is such that the least numerous sex tends to determine the effective population size. For most livestock breeds this means males. The most obvious consequence of this is that effective conservation breeding usually involves using more males than would be absolutely necessary if the only consideration were the number of females a male can manage to mate. Genetic consequences dictate a different answer for number of males than do husbandry considerations. As an example, a population of 50 animals could be 40 females and 10 males, with an effective population size of 32. If the population were 25 males and 25 females, then the effective population size is 50. Constraining the population to have relatively more males raises the effective population size dramatically.

74 Maintaining Breeds

Number of Males	Number of Females	Effective Population Size
25	25	50
10	40	32
5	45	18
1	49	4

Consequences of population sex ratio on effective population size.

The equation given above for effective population size is indeed oversimplified, and a longer but more accurate equation includes the interval from parent to offspring for each sex, and the degree of relationship between animals. These are such important issues for effective population size that it turns out the Holstein cattle breed with a global census of millions has an effective population size of about 30 individuals. This limited size makes significant inbreeding impossible to avoid. The result is that inbreeding is having an effect on viability and reproduction in this breed. Unfortunately for breeders of rare breeds, it is impossible to capture all of the information for an accurate determination of effective population size in a relatively easy manner. The level of detail needed to arrive at an accurate answer is overwhelming. The simplified equation is useful in pointing to the importance of sex ratio, especially in rare breeds, but it is important to note that the number generated by this result is likely to be a relatively optimistic one, and that true effective population size is generally even lower than predicted by the simplified equation.

The key points to remember are that effective population size is generally lower than the census size would suggest, and that any shared ancestry (especially in recent generations) among breeding partners takes that number still lower. Especially for rare breeds the effective breeding population size is likely to be much smaller than the outright census, and this is a more critical factor for breed survival in rare breeds than it is for breeds with large populations. This is because inbreeding in rare breeds is more likely to be widespread throughout the population. The principles of effective population size are a compelling reason to keep track of different bloodlines within rare breeds so that breeders can assure that outcrosses are available to each and every animal in the breed. No specific guidelines can be offered to suggest a ratio between census and effective population size, as this is a complicated biological concept and oversimplifying it leads to a false sense of security about the true genetic composition of purebred populations. Breeders

Effective population size is nearly always lower than census size.

should monitor their breeds closely to assure that outcrosses are available within the breed, and should be diligent to keep different bloodlines viable and secure as insurance against future needs.

Generation Interval

Generation interval refers to the average age of parents at the time offspring are born. It is not the age of the parents at the time of the first offspring, but is instead the average age of all parents and for all offspring. As a result it is a difficult computation for many populations. A simplified example is that if sows and boars are used to produce only a single litter, then the generation interval is about a year. If, in contrast, sows are kept up to seven years and boars for five, then the generation interval goes up to about three and a half years for the population (assuming an even distribution of ages up to the maximum age).

Generation interval is the average age of parents at the time offspring are born.

Selection programs to promote genetic improvement generally favor as short a generation interval as is possible. This allows breeders to maximize genetic gains per year, for (ideally) each generation should be genetically superior to preceding generations if selection is indeed successful. Decreasing the interval between generations should therefore hasten genetic improvement. It is true, though, that genetic superiority may not always result in production superiority. For example, a genetically superior two-year-old dairy cow is unlikely to produce as much milk as a somewhat more genetically average six-year-old dairy cow because the six-year-old is benefiting from being able to produce her mature potential. In this situation the increased production potential coming from genetic gain can only be fully realized in the milk pail if the cows are kept to the age at which mature production is realized.

For many rare breeds the generation interval may well need to be kept longer than is typical for industrial breeds. One reason for this is that older parents have demonstrated their adaptability and production success over longer time periods than have younger ones. They are therefore safer bets for having the genetic strength desired in an adapted breed than are parents that are younger and less well proven in the environment. In addition, especially in small populations, increases in inbreeding are related to generation interval. Keeping the generation interval as long as possible is one tactic to minimize a dramatic annual accumulation of inbreeding, because any inbreeding is accomplished over a much longer period of years as the generation interval stretches out from the use of older animals.

Especially in the case of foundation animals, the wisest plan is to use them

This elderly Choctaw mare has proven her strength by remaining a sound mount as well as producing foals for over twenty years. Such performance cannot be documented in younger animals. Photo by D. P. Sponenberg.

until they are no longer reproducing. This increases the opportunity for them to contribute their entire genome to the breed. Exceptions to this rule are important, and include situations in which few founders in a small breed may swamp the entire breed by becoming bottlenecks if all replacements are saved from these few. In that instance it is best to balance the contributions of all population members by tightly constrained breeding programs.

Males are much more likely to over-contribute their genetic influence if kept actively breeding for long careers, because they produce more offspring than do females. In many rare breeds or rare strains, a relatively common history is that a herd used a single sire at a time for multiple years, replacing him with a son. Randall cattle and Conway Pineywoods cattle fit this model. The initial result of using the male is the production of daughters and sons, but with retention of daughters in the herd the male is relatively quickly being mated to daughters as well as to the original females. If he lasts a long time, then he is likely to be mated to his granddaughters that were produced by daughters. When he is eventually replaced by a son, this son is related to the entire herd. Following several cycles of this sort of breeding all animals in the population are likely to be very closely related. This situation can be exacerbated with the widespread use of artificial insemination. One tactic that can be used to avoid the males becoming a genetic bottleneck

Foundation animals should be used wisely, but for as long as biologically possible.

is to replace the males relatively rapidly, while maintaining the females for as many years as is practical. Keeping both sexes for long reproductive lives tends to lead to a genetic bottleneck, but a drastic reduction in the reproductive lives of both sexes likewise tends to lead to constricted genetic pool due to a very short generation interval. Having one sex at a short interval and the other at a very long interval is a good compromise for the long-term genetic health of the population.

Inbreeding within Individual Herds

Degrees of inbreeding levels are likely to be higher within an individual herd than they are across an entire breed. This is largely due to the use of relatively few males in any herd, with retention of breeding stock from a single or only a few sires. Over several generations of closed-herd breeding this strategy can lead to fairly significant inbreeding within a herd. This can indeed be one of the sources of the uniformity that is usually desired in a herd, but must be done with forethought.

In most breeds, inbreeding within a herd is little threat to the overall genetic health of the breed. This is true because other herds are either being outbred or are being inbred to different individuals. As long as the inbreeding is not the result of the same few individuals across the entire breed, then the breed is relatively safe and is probably not losing much genetic variation. The reason for this is that one inbred line can be linecrossed to a second but unrelated inbred line, and all of the inbreeding that is built up vanishes in the next (linecrossed) generation. However, a backcross to either parental line will assure that inbreeding is once again occurring, so wisdom and close management are needed if the goal is continued assurance that outcrosses are available to every animal in the breed.

Inbreeding can be one source of uniformity.

Inbreeding within Breeds

Inbreeding is much more problematic when it is uniformly distributed across an entire breed rather than within separate herds. When inbreeding is uniform at the breed level it is nearly impossible to avoid its consequences. If all breeders pursue the same successful bloodlines and individual animals, then the result is that the entire breed is being inbred in the same direction. This can be visualized as a circle with the tension at the boundary all being directed into the same middle point, which is usually a popular individual animal or bloodline. Over several generations the circle, which represents the overall genetic variation in the breed, becomes smaller and smaller. Popular sires or bloodlines that expand

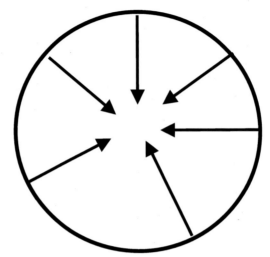

Inbreeding in a single direction yields a constricting gene pool for the entire breed.

at the expense of others in the breed are common culprits for significant inbreeding problems across entire breeds. It is especially dangerous to use uncontrolled artificial insemination, as the entire breed can be mated to a handful of related sires. Such a strategy accelerates the inbreeding that can occur.

The ever-contracting inbred circle can be contrasted to the situation in which different bloodlines within the breed are being bred in different directions and to different individual animals. In that situation the inward forces are subdivided to subcircles within the larger breed circle. The result is that the overall breed circle tends to maintain its original size without much loss. Genetic variation is maintained, and with it, breed health.

Strategies for managing inbreeding vary from breed to breed. At one extreme there is no control at all over inbreeding, nor strategies for managing it. Some dog breeds, and cattle breeds such as the Holstein, are among those that allow for breeding to be concentrated on a very few individuals. The result is unavoidable inbreeding, and some of these breeds are now reaping the unfortunate consequences of this as diminished vigor. The philosophy and practice of purebred breeding, as envisioned over the last two centuries, have not yet fashioned a good strategy for dealing with inbreeding in closed populations. Over the next few decades, breeders and associations need to address this issue if the genetic heritage of breeds is not to be lost through their attrition to extinction from lack of vigor.

Some breed associations may well choose to control inbreeding by limiting the number of offspring registered per animal per year. This can be especially

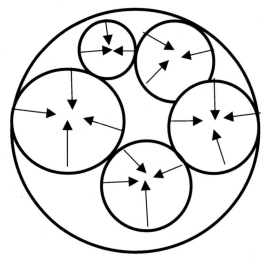

Inbreeding in different directions can keep more genetic diversity in a breed.

important (if rarely done) in breeds in which artificial insemination and embryo transfer are allowed. Putting a limit on the number of offspring an animal can produce in a year or a lifetime assures that no single animal swamps the entire breed. This can be a very difficult step for a breed association to take, as it pits the long-term survival of the breed against the short-term economic benefit for individual breeders who have popular animals.

In many breeds it may not be necessary to have a formal set of rules to manage inbreeding. In most breeds it is possible to educate breeders about the bloodlines within the breed and the need to keep these going into the future. A quick analysis by the breed secretary can usually reveal which lines are becoming more rare and which are in danger of swamping the breed. Alerting the membership to the status of the bloodlines can boost activity within the rarer bloodlines, especially if this is matched with long-term educational endeavors on the part of the breed association about the importance of bloodlines within a breed.

Strategies for managing inbreeding vary from breed to breed.

Combining Linebreeding and Linecrossing

One mechanism for population maintenance that can work well for a number of breeds uses the advantages of both linebreeding as well as linecrossing. Matings can be constrained so that following a linebred generation the animals are then linecrossed. By alternating the two mating strategies generation to generation it is possible to reap the benefits of each without experiencing too many of the negative aspects of either. Such a plan does require great attention to detail, because matings must always consider the fate of the animals several generations into the future.

Inbreeding and Outbreeding Summary

Both inbreeding and outbreeding have a place in breed and herd maintenance. Making sense of all of the details of inbreeding and linebreeding can be perplexing, especially when these issues hit the barnyard and the breeding decisions that must be made by breeders. Associations can help their breeders by educating them concerning the lines within the breed and the importance of these to breed-wide genetic health.

Especially difficult is the very real possibility that short-term economic interest frequently places pressure on breeders to all go in a single breeding direction by trying to produce animals that fit current demand. In contrast, long-term breed viability requires that breeders be maintaining sufficient diversity for breed viability. Managing the tension of these two demands on breeders is one very important role for associations and breeder communities.

Over-representation of Individual Animals

In an earlier section it was stressed that every animal should have a potential linecross within the breed to assure the breed's genetic health. This requires attention and careful planning, otherwise all animals of a breed can become related to one another. This usually occurs through the over-use of individual excellent animals. For some breeds of livestock, certain individual animals have become over-represented in the breed, to the extent that it is nearly impossible in some breeds to find linecrosses. When certain animals become over-represented, other animals become under-represented, and so the breed loses their genetic contribution. When individuals become over-represented it means that the breed has lost variability, and could be in danger of losing the genetic variation essential for long-term survival.

Individual animals can become over-represented; consequently other individuals then become under-represented.

One way that individual animals become over-represented is through the "founder effect". Founder effect describes the phenomenon that a population descended from a few founders cannot contain any genetic variability that is not contained in those founders. Founder effect is very constricted when new herds are started with a few animals, and is especially the case when only a single male is present in the newly established population. It is not unusual for an original founder male to account for a great proportion of the genes of a population, largely because he and then his sons were used in the group for generation after generation. This is especially likely in traditional systems where a popular male might well be used until he physically plays out. At this point he has usually swamped the population with his own offspring, and has been mated to his own daughters, thereby increasing his genetic impact on the population all the more.

The Randall (or Randall Lineback) cattle breed provides an example of founder effect. This breed now descends from a limited number of founders, all of which were closely related and all of which were rescued from extinction by Cynthia Creech and Phil Lang. Detailed pedigree records were, unfortunately, not available at the time of the breed rescue. The most optimistic analysis, though, still has all cattle related to a single male, whose contribution to individuals of the breed ranges from 12% to 48%. In this case it is impossible to avoid inbreeding to that one founder, though careful breeding management can assure that relatively low percentage cattle are always in the population. The low percentage cattle allow for the relatively high percentage cattle to be mated strategically to avoid concentrating that one founder. The other founders in the Randall breed vary in their contribution to individual cattle from 50% to 0%. The variation in contribution allows for some distance in matings between cattle so that higher percentage cattle can be mated to lower percentage cattle for the various founders. If breeding during the rescue had not been carefully monitored it could easily have occurred that the contribution of certain founders would have been uniform across the breed, at which point all matings would be linebred.

Founder	Average % in cattle alive in 2005	Minimum % in cattle alive in 2005	Maximum % in cattle alive in 2005
1	33	12	48
2	16	0	50
3	15	2	25
4	5.7	0	38
5	7	0	25
6	4	0	50
7	5.9	0	25
8	2.5	0	19
9	4.6	0	25
10	2.4	0	13
11	2.1	0	25
12	1.5	0	50

Management of percentage contributions of founder animals through breeding management of Randall cattle.

Randall cattle all trace back to a very small founder population. Photo by D. P. Sponenberg.

One successful strategy to avoid over-representation of a founder male is to use males for only a single breeding season or maybe for two. In this way a foundation sire's sons are used on foundation females, instead of the original foundation sire being used his own daughters. The resulting progeny are then only 25% his influence, rather than the 50% that his direct sons and daughters would be – and much less than the 75% that would result from mating him to his daughters. Though the son will also be mated to his sisters to result in 50% relationship to the original sire, this is still much less than the 75% result of using the original sire on his daughters. The strategy of using males for a short time allows for the females to have a much higher proportionate effect in the population, and is a useful strategy for assuring more uniform genetic contributions of founders.

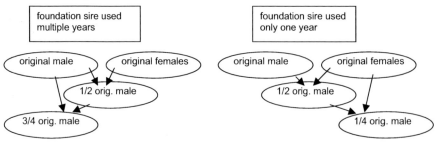

Foundation sires can be used widely or sparingly, and each strategy results in a very different outcome for the population.

Another way that individuals become over-represented is the "popular sire" phenomenon. The popularity of a given male may be such that many breeders use him, or his sons. This can easily swamp a breed, and is a major problem in some purebred dog populations. Show ring success is commonly the underlying reason for a sire's popularity. In many cases breeders have discovered only in later generations that a genetic problem has been identified with some such popular sires – and by that time eliminating the problem has become a major logistical headache because of his influence throughout the breed. A much more secure path is to make slow progress by assuring that no single animal becomes over-represented in a breed.

Genetic Bottlenecks

Genetic bottlenecks occur when only a few individuals of a breed remain. Bottlenecks are a drastic form of over-representation of specific animals. These individuals then become the founders for the entire future of the breed. The bottleneck concept can be important in deciding the fate of breeds. For example, some breeds such as the Texas Longhorn and the Colonial Spanish Horse once numbered in the millions. Populations then crashed to several hundreds, and the registries were founded on still fewer individuals. The contemporary breed population goes back only to those individual animals that founded the registered populations, so the bottleneck that they represent now provides the total genetic variation in the entire breed.

Bottlenecks reduce the genetic variation that should provide breed definition and predictability and they constrain viability in some situations. A "short" bottleneck, such as a single constricted generation of low numbers followed by rapid expansion, is much less harmful than a "long" bottleneck. Long bottlenecks, especially when only a few males are used for several generations in a row are especially damaging. Cloning to rescue an endangered breed, such as occurred with the Enderby Island cattle, is a sort of ultimate bottleneck.

In any breed, a severely limited male population constrains the next crop of offspring to half siblings, and the result over several years can be to severely limit genetic variation. This has worked in some situations to provide serviceable, viable, and productive breeds such as Randall cattle, Conway Pineywoods cattle, and Farceur Belgian horses. These success stories can lull breeders into thinking that bottlenecks are of no consequence, but in most populations such an event leads to inbreeding depression and failure of the population. Bottlenecks must be managed carefully so that purebred outcrosses are always available.

Artificial Insemination, Embryo Transfer, Assisted Reproductive Technologies

A variety of assisted reproductive technologies have emerged over the last decades, including artificial insemination, embryo transfer, embryo splitting, as well as the artificial incubation of poultry eggs. These technologies are powerful, and must be used wisely in purebred breeding programs. For most of these, the tendency is to use them to more widely use an outstanding individual than would otherwise be the case through natural reproduction. This strategy usually results in an over-representation of that outstanding individual.

While the assisted reproductive technologies do tend to lead to genetic bottlenecks, they do not necessarily lead in that direction. If used wisely and for targeted goals, they can be very helpful in managing rare breeds to have genetically sound and viable structures. For example, semen and embryos can be saved from foundation animals so that their genomes are present for future use. Going back and occasionally using semen stored from long-dead ancestors to the contemporary breed can also lengthen generation intervals. Additionally, important and genetically under-represented females can be superovulated in order to assure broader contribution to the breed. It is important to recognize that the technologies are in themselves somewhat "value neutral" being neither bad nor good for breed maintenance. It is the specific way in which they are used that determines whether they work for the positive benefit of breed genetic security. They can be very powerful in making positive contributions, although the fact remains that by their very nature of expanding possibilities of enhanced contribution of single individual animals, they tend to usually work in the opposite direction of decreasing genetic variability and long-term viability of breeds.

Mating Systems and Pairing of Animals Within Purebred Breeds

"Mating system" refers to the specific manner in which males and females are brought together. Choices include individual pairing, group mating, and multi-sire systems. Mating systems have important consequences for pedigree maintenance, and most purebred breeders put great importance on pedigrees. A pedigree is an account of the ancestry of an animal, and identifies specific individuals as sire, dam, grandsire and so on. That is, pedigrees link an individual animal to certain ancestors. Pedigrees are of great value in establishing the kinships of animals within a breed so that inbreeding can be monitored. Pedigrees are useful when choosing mates for the next generation for whatever breeding strategy is deemed appropriate.

Individual pairing is the mating of one male and one female at a time. In most situations this is usually limited to horses, donkeys, and dairy goats. Most of these matings are handmatings, where the animals are haltered and controlled, and the animals are never turned out as a breeding group. This situation can also be mirrored by geese, which form such strong pair bonds that matings will invariably be between bonded pairs even if the geese are run in groups of multiple pairs or trios of birds. One strategy for managing geese is to pair (or trio) them for their first breeding year, and then in subsequent years put them in larger groups. The geese still mate in the original pairs or trios, but the result of the larger groups is easier management, with no loss of pedigree information as long as eggs are identified as to the goose producing them. Most ganders in a trio will have a primary and secondary mate, and as the years progress, they often begin to ignore the secondary mate so that her fertility may fall. Diligent breeders pay attention to this so that productivity can be monitored. Goose pairs can be disrupted, but usually this requires pairing a goose with a new single gander and assuring that the previous mate is out of contact range.

"Mating system" is the manner in which males and females are brought together.

Single-sire group matings are more common for most domesticated species. This situation uses one male to mate a group of several females. The sex ratio depends on the species, and can be as high as one male to fifty females (sheep, goats) or as few as one or two females (geese). For pedigree management some producers use different males sequentially over a group, keeping track of dates that each specific male is in the herd so that accurate pedigree information is maintained.

Both individual mating and single-sire group mating provide for accurate pedigree information based on good record keeping. With these two systems it is not necessary to resort to the expense of DNA testing to assign

African geese are typical geese in forming strong pair bonds. Photo by D. P. Sponenberg.

parentage, although some breed associations still require this as a further check on the accuracy of registry information.

The high regard given to pedigrees by most animal breeders needs to be viewed from the perspective of breed history. Many breeds of conservation interest, including several that are among the most phenotypically variable and regionally adapted, came to the present day through a long period as landraces. Mating of landrace animals in traditional settings was and still is commonly accomplished in multi-sire herds. Therefore, accurate pedigrees on the sire's side are not possible unless breeders resort to DNA typing or bloodtyping. Such documentation is generally only worthwhile in species with high individual monetary value such as cattle and horses. It becomes uneconomic for animals having relatively low individual monetary value such as chickens, turkeys, ducks, as well as most sheep and goats. Rabbits and swine avoid this issue because these species rarely use multi-sire mating systems.

The high regard given to pedigrees needs to be put into perspective.

For many adapted breeds, and even for several of the more highly selected production breeds, multi-sire herds can offer real advantages for breed maintenance. The multi-sire tactic for breed maintenance should not be dismissed simply because it negates the possibility of pedigrees. At a minimum, multi-sire herds put selection pressure on male competition and survival fitness. This can be especially important in primitive and adapted breeds, where competition is the only means to assure that the fittest are indeed surviving and reproducing.

Heritage turkeys such as the Jersey Buff, when bred on the range, rarely have accurate pedigree information. Photo by D. P. Sponenberg.

Some species, even if managed in multi-sire situations, can have accurate pedigree information. Depending on numbers and range, for example, horses tend to form relatively stable harems, and running multiple stallions on a single range has been a strategy long used by breeders of Choctaw and other Colonial Spanish horses. By noting the harem membership it is possible to note which stallions sire which foals. Other species form harems poorly, so that this strategy does not work for cattle, sheep, or goats.

It is possible to gain some benefits from multi-sire mating systems without completely losing the benefits of pedigrees. One strategy could be to assure that the sires in the herd at any one time are all of the same bloodline. While pedigrees would not be available on the offspring, their status as being of one bloodline or another would be available. This information could be very helpful in managing genetic diversity across the herd or breed. An example of this is using a group of three Hickman strain bulls in a Pineywoods cattle herd, followed by a group of three Ladner strain bulls. The rotation of the bulls as a group provides for knowing that the calves have a certain strain of sire, and this helps to place the calves into genetic groups.

Another strategy is to use males sequentially in the herd, one at a time. This can be unrealistic in large herds or on large ranges, but can provide for pedigree information. If males can be used sequentially, then only a few individuals that are born during transitions may be difficult to identify as to pedigree. Allowing for a break between males can circumvent this problem. For example, a large group of doe goats could have an individual buck in for a week, then none for a day or two, then a second buck for two weeks, none for a day or two, and finally a third buck for two weeks. The result is a five-week kidding period during which the kids come from three different sires and are all accurately identified as to sire. The advantage of this approach is that multiple pastures are not needed; the disadvantage is that no specific doe is guaranteed to mate to the most preferred buck for her.

Selection

Selection is important in maintaining pure breeds, regardless of the mating system and the breeding strategy used by breeds and breeders. Selection refers to the inclusion of specific animals as breeders, and the exclusion of others. The result is that some animals reproduce, and others do not. Those that do not reproduce can still have important roles in fiber and food production, as pets, or for draft power. All of these can be important for breed promotion. Among those animals that do reproduce, each individual's relative contribution to the next generation can vary considerably.

Selection is a powerful tool for changing or maintaining an animal population, and is a critically important aspect of breed maintenance. Selection is a very important source of uniformity within breeds and herds. Any individual breeder tends to favor a certain look and function over others. Over time a herd under a single breeder's care and selection will tend to express that look more and more strongly. Selection and inbreeding frequently act together as potent forces for uniformity especially in individual herds.

Selection assures that the more desirable animals produce more offspring than do the less desirable animals. The definition of "more desirable" and "less desirable" varies with situation and philosophy, and from one breeder to the next. Differences in selection can have profound repercussions on the breed and its future. While understanding that selection is one aspect of breed improvement, selection is also important for breed maintenance.

Selection is important in maintaining pure breeds.

Selection shapes future generations of a population because certain genotypes are included and others are excluded from participation. Selection is a powerful tool that can work to strengthen or to weaken the status of breeds as genetic resources.

Variability in selection has caused differences in the style of animals produced by different breeding programs. The traditional ox-using Pineywoods cattle breeders insisted on horned cattle. Some breeders more interested in beef production have favored polled animals. Conway strain cattle, as a result of selection, are nearly all red and white in some combination. Hickman cattle have no such color theme, as selection favored color variation. Some breeders liked and kept a few guinea dwarf cattle in their herds, others avoided these and the result is absence of guineas in some strains.

Degree of Selection

The degree of selection that occurs in a population depends on the relative number of animals used for reproduction. For some populations the degree of selection, especially for females, is fairly low. That is, nearly every female is used for reproduction and therefore gets a chance to pass along her genes for good or ill. In essence this means that little selection is occurring on the female side of the equation. This situation is typical of breeds or species that are involved in the novelty market or are newly present in the market place such as recently imported breeds.

For most breeds the degree of selection for males is higher than that for females, but how much higher varies with individual breeds as well as with individual breeding programs. In some herds nearly every male that is born is used for reproduction, while for other herds only a small percentage have the opportunity to reproduce. The overall result is that the degree of selection for males is indeed higher than for females, and varies from moderate to extreme. This can be good or bad, depending on how selection is used.

An example in dairy cattle illustrates the degree to which selection can proceed. The degree of selection for dairy bulls (especially Holsteins) is probably the highest of the common domesticated species, because by artificial insemination only one bull in thousands is used. Over several generations this has profound consequences for the genetic composition of a breed. In each new generation, the bulls selected for semen production are nearly invariably the offspring of a previously selected semen sire. The result here is that the new sire is related to many of the females of the breed (his half-sisters) but also to several of the males (his half brothers). A trend to save bulls out of certain cow families (half-sisters from specific sires) also contributes to concentration of the genetics as coming from a very few individuals. This, over generations, compounds the narrowing of the genetic base. The intense degree of selection and the resulting narrowed genetic base have long-term consequences for the overall genetic health of the breed. In contrast, the relatively low degree of selection for many alpaca herds (many males see at least some use, with only a very few used widely) indicates that more genetic change is easily possible than is currently being realized, simply

The degree of selection affects the relative genetic superiority of animals that reproduce.

In the Holstein cattle breed, only one bull in thousands is used for reproduction by way of artificial insemination.

Few alpaca males in North America see wide use on maximum numbers of females. Photo by D. P. Sponenberg.

by increasing the degree of selection used among the males.

To effectively manage the genetic component of breeds, and especially numerically small breeds, it is generally wise to not have a degree of selection as high as is typical of main-stream industrial dairy cattle. Progress for genetic potential for production can be rapid under such high degrees of selection, but the disadvantage of this tactic is that a great deal of genetic material is discarded in unused males. The result can easily be very restricted genetic variation that approaches nonviability, especially if such high degrees of selection are used for several generations in a row. The other extreme, that of using each and every animal for reproduction, assures that no improvement in performance, and no penalty for nonperformance, is achieved in the population. Both extremes work to the detriment of a breed, and the best strategy usually lies somewhere in between these extremes.

Turkeys provide both extremes of selection. At the industrial end, turkeys are the result of an extraordinarily high degree of selection and are very genetically uniform. In contrast, many back-yard breeders have very limited selection and may use a tom on only one or two hens. Heritage turkeys of the standardized varieties that were once commercial have a genetic background that included mating of toms to large groups of hens (about 12) so that degrees of selection were adequate to assure progress in commercial utility of this important type of turkey. Fortunately, more and more breeders are learning the techniques of selection that can be applied to this type of bird, and their important historic role is being recaptured.

Selection and Type

Selection is a sword with two edges. Selection is a powerful tool for change, but not all change is good. Loss of a breed's genetic heritage is obvious when it occurs through outright extinction of the breed, but a more insidious loss involves deviation from the original type of the breed so that it no longer represents the original gene pool. This second kind of loss can occur rapidly and completely if crossbreeding is occurring. A similar loss occurs, if more slowly, by selection within the breed away from the traditional type or form of the breed. As a result of recent selection pressures, many mainstream beef cattle breeds are all remarkably similar in type. A can of spray paint could render them impossible to distinguish by breed. One underlying philosophy of breed conservation is that the original form of breeds is important to keep so that choices remain available to future agriculturalists.

The original form of breeds is important so that future agriculturalists have choices.

To preserve breed type it is important for breeders to constantly select animals that reflect that type, and to reject those that deviate from the original breed type. Some of the South American literature on the Criollo horse is instructive in this regard. In the mid-1900s selection began to increase the animals from their traditional height of 14.2 hands. With that increase in height came a perception of great beauty and eye appeal, but much less athletic prowess. Breeders

This Argentine Criollo is an excellent example of the traditional utility type that is once again favored by breeders after they briefly favored taller horses. Photo © Evelyn Simak.

wisely abandoned the quest for height in the breed, and returned the breed to its original form before irreparable loss had occurred.

Other examples of changes in breed type are numerous. Today's Morgan horses vary from the original multipurpose farm horse to a very specialized park-type show horse. Many hog breeds have varied from very obese lard types (not in much current demand) to very lean – all within the same breeds as they have been changed by selection over decades for a new and different ideal type. Angus cattle have changed from low, wide, fat cattle to much taller, rangier, muscular cattle over the last fifty years.

Controversies over type have no easy answers, but maintaining type is fundamental if conservation is to truly save the unique genetic packages that breeds are. Type is essential to breeds, and must be closely safeguarded rather than allowed to drift. Breeders must be able to understand and appreciate type for this to occur. Guarding type can become an extremely difficult issue in some breeds where historic types (lard type hogs, dual purpose cattle, farm horses) meet with little current demand, while changes in type can help them to more successfully compete in the modern marketplace. These are complex issues with no easy remedies, because guarding types that are undesired in the contemporary marketplace nearly assures relegating breeds to rarity.

Selection of Animals for Reproduction

Breed type, performance, and selection issues bring into question just exactly which animals should be used for reproduction, and which should be culled. The goal of most breeders of purebred livestock is to constantly improve their stock for their specific goals, while staying within the constraints of breed type and purpose.

Most successful animal breeders have an "eye," or an intuitive sense, for which matings will work and which will be less successful. This is somewhat at variance with a more quantitative and statistical approach that has been followed by scientific animal breeders over the last several decades. Both approaches have merit, but the eye of the master breeder has, and still does produce excellence both in performance as well as in eye appeal. Most breeds benefit from a combination of the more statistical animal breeding approach as well as from the "eye of the breeder" approach.

Matings for stock improvement take the form of balancing weaknesses with strengths. For example, the weaknesses of one parent (set of hocks, for example) might be balanced by a mate with stronger, truer hock conformation. Or, a productive and useful animal with an off-type ear could be mated to a more "typey" mate to try to correct the type faults while maintaining production. Note well, though, that only rarely is a fault corrected by mating to the opposite fault – such a strategy is usually very disappointing. So, an animal with sickle hocks is rarely corrected by mating to a post-legged animal, but is rather more likely to be corrected by one with a sound set of hocks. Balancing weaknesses with the ideal is a much surer path to success than is attempting to balance a weakness with an opposite weakness.

Not every mating in every herd has the same goal or tactic. Understanding this helps breeders to shape their herds. Each animal within a single herd has a different importance in respect to its genetics, productivity, and conformation. In a very real sense, each animal in a herd has its own role and potential, and realizing this is key to making genetic progress at the population level. Likewise, each animal and each mating have different importance to the overall breed. For example, a sole female of a unique foundation strain needs to be used much more wisely than a female of a more numerous composite herd. Likewise, a sole male of a rare strain might be allowed minor faults to a greater degree than a male of a common strain in order to assure that the positive attributes of the rare strain are not irretrievably lost.

Matings for improvement target balancing weaknesses with strengths.

The planned and hoped-for outcomes of certain specific matings or certain specific individuals within the herd are likely to be different. Certain matings target the production of future breeding males, while others do not. The produc-

A typey Choctaw stallion can be mated to mares of varying strengths to produce foals with different uses in conservation programs. Photo by D. P. Sponenberg.

tion of breeding males ideally comes from mating a very typey and productive female to a typey, productive male. In contrast, less typey or less productive animals are usually mated to superior mates in the hopes of offspring that correct some of the parental faults. A male offspring from such a mating that compensates for weakness may well not find use as a breeding animal so that the weaknesses present in the one parent are not widely spread in the population. Or, occasionally, a male from such a mating may see limited use on select typey, productive females to produce strong and balanced offspring that consolidate more strength and less weakness before this genetic material can be more widely used in the next generation. In contrast, the female offspring of a relatively weak parent may be an improvement over that parent, and may be a safe avenue to assure that the genetic contribution of that original parent to the breed is not lost. Female offspring generally pose less risk of overwhelming the breed through their breeding careers than do males, so that animals must not only be evaluated by their own performance but also by that of their ancestors in an attempt to assure that only strong, productive genetic material is widely spread throughout the breed.

Each mating should consider the role that potential offspring will have in the breeding population.

Each mating should have a goal of improvement and genetic strength, and each should also be evaluated as to how the offspring will fit into the population as a potential breeding animals. That is, the main mindset to use when evaluating breeding animals is to figure out just how their offspring (male and female) will fit into the population and be used. A typey and strong female, for example, can be mated to a typey and strong male for male and female offspring that see wide use. The same female, mated to a somewhat weaker male, may provide female offspring to be used for specific purposes, but not males that are to see wide use. One reason for such matings might be to balance bloodline representation in a herd, but nonetheless carries with it expectations for the use of the offspring that are produced. Animals that are typey and productive are mated to one another for the production of offspring that will see wide use (generally males), while animals of less strong type, production, or ability may see more limited use or may indeed be culled from the breeding program.

Use of Estimated Breeding Values

In many commercial breeds it is common to have statistically-derived estimates of the genetic worth of individual breeding animals. These are known by different names in different species and in different breeds, but common ones include estimated breeding value (EBV) and estimated progeny difference (EPD). These estimates are derived from the performance of relatives and progeny. These techniques usually involve the measurement of commercially important traits such as growth rates, milk production, reproduction, and the like. Associated with each EPD is usually a repeatability score, which is a reflection of the accuracy of the estimation. The more family members that are measured, the higher the repeatability on any individual. The repeatability can be very important as a guide to the expected outcome of matings, and is usually highest for old, proven sires than for any other age or sex class.

The EPD values are calculated against breed averages, and are in the units in which the trait is measured. For example, in sheep it is possible to generate EPDs for the prolificacy of a ram's daughters. These might run from 0 (breed average) to –0.3 (daughters less prolific than breed average by about 0.3 lambs per daughter) or + 0.5 (daughters producing about half a lamb more than the breed average). How to use this information is up to the breeder. If the breed is relatively prolific, it might well be that in some situations the best bet is the ram with the EPD of – 0.3, because the goal might be try to reduce the number of triplets and quadruplets. In contrast, a different farm with a different physical set up, and different goals, might well want the +0.5 ram, because the farm and management can successfully manage and benefit from all those extra lambs.

A second example might be birth weights, in fractions of a pound. EPD values of 0 (breed average), +1 (one pound heavier at birth) and –0.5 (half a pound

This Jersey cow represents mainstream commercial dairy selection, which has used EPD data to constantly improve milk production levels. Photo by D. P. Sponenberg.

lighter at birth) could all be used to advantage in different situations. EPDs can be calculated for a number of traits, depending on what the breeders choose as important targets of selection programs.

The EPD statistics have great value, and can be used different ways. In production herds the underlying philosophy is usually one of improvement, which might not always be appropriate if the goal is to maintain animals adapted to harsh environments. Improvement means change, and not all changes are desirable in all situations. In herds where certain traits are at an acceptable level, then using EPD-rated sires with values of "0" may be an informed and targeted decision.

Gene Flow Into and Out of Breeds

Genes can and do flow into breeds from outside sources. Crossbreeding to other breeds is the usual source of gene flow. Some of the flow of genes into pure breeds is inadvertent, but some inflow is due to short-term goals and fraud on the part of breeders. Gene flow into breeds, especially if widespread, can easily corrupt and permanently change the breed package to the point that it is no longer truly a genetic resource.

There are many examples of crossbreeding which has led to loss of breed character. Among these examples is the current convergence of nearly all black-faced sheep breeds in the USA onto a very Suffolk-like frame and head type. A long-time Hampshire sheep breeder interested in pure breeding related in the 1990s how he had become disgusted by the fad towards very Suffolk-appearing Hampshires, and decided to leave the Hampshire breed. But first he did an experiment by crossing Suffolks to his remaining Hampshires. The first crosses were very close in type to traditional Hampshires, thereby indicating that the current fad in "Hampshires" was likely at least 3/4 Suffolk breeding.

To remain a viable genetic resource, breeds must be kept pure. This is especially challenging for landraces such as Spanish goats. Photo by D. P. Sponenberg.

Maintaining Breeds

This crossing, especially if reinforced by show ring success or other market forces, is the death-knell for the genetic uniqueness of breeds. With that loss also goes the true usefulness of the breed, as it becomes a somewhat second-rate copy of a different breed.

The Hampshire sheep example demonstrates that show ring fads that do not place emphasis on breed type can quickly spell the end of a breed as a purebred genetic resource. Once crossbreds are acknowledged as purebreds within a registry it is difficult or impossible to regain the genetic package that is so essential to the purebred's ultimate function. The show ring can be particularly damaging to breeds if off-type animals are placed high in the rankings. This provides an incentive for breeders to abandon selection for breed type, and can easily lead to introgression of crossbreds into a pure breed.

Show ring fads can be the end of a breed as a purebred genetic resource.

Upgrading and What It Does

Upgrading is the sequential use of purebred animals over a series of generations to provide a "nearly purebred" result. The usual sequence is that a purebred sire is used on females that are not of his breed. These females can either have undocumented ancestry, be crossbred, or be of another breed. The resulting offspring are 1/2 (50%) the pure breed of the sire. The daughters are then mated back to another purebred sire of the breed, providing offspring that are 3/4 (75%) the pure breed. The next generation provides offspring that are 7/8 (88%); the next is 15/16 (94%), then 31/32 (97%), 63/64 (98%), and 127/128 (over 99%) and so on. These part-bred animals are called "grades" of the pure

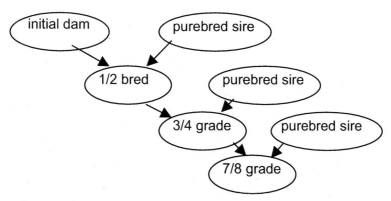

The use of a sequence of purebred sires provides for offspring of increasing percentage breeding of the pure breed. This is the essence of upgrading.

breed to which they are being sequentially mated, and many breed associations designate a specific percentage at which the level of grade is designated as part of the purebred population.

Upgrading in most breeds involves only the use of purebred sires on grade dams. It is also biologically valid, if rarely done, to use purebred dams for this process. It is also worth pondering that due to the unique genetic contributions of females (mitochondrial DNA, for example) it may actually be best for upgrading programs to insist on the use of both purebred males and purebred females in some of the generations to assure that breed-appropriate genetic material that is sex specific (Y chromosomes from males, mitochondria from females) has come from the purebred pool.

Upgrading is the sequential use of purebred animals over a series of generations.

Upgrading has a legitimate and important role in some breeds (notably landraces or recently developed breeds) but is generally considered to have little if any benefit to older, long-established breeds. The truth is that upgraded animals offer nearly all pure breeds a real opportunity for vitality and viability without endangering the breed's genetic heritage. Upgrading allows breeds to avoid the problems of a completely closed gene pool with its attendant inevitable inbreeding, while at the same time safeguarding the status of breeds as predictable genetic packages. Upgrading is a way to ensure breed survivability over long centuries of purebred breeding. Understanding what upgrading is and how it affects a breed genetically are extremely important issues to breeders of all purebred breeds, whether or not they choose to use this strategy.

Upgrading has been widely used in several livestock species, especially with recently imported production breeds. It allows for rapid numerical expansion of

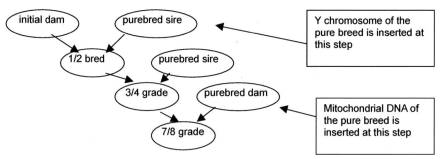

By using both males and females from the pure breed involved in upgrading it is possible to assure that both the mitochondria (female) and Y chromosome (male) come from the pure breed.

the breed, and also provides a demand for purebred animals (generally males) for crossbreeding and upgrading purposes. Upgrading must be managed intelligently, though, or it becomes an avenue for the inclusion of animals into purebred herd books before those animals have acquired the genetic package that makes the breed distinctive and useful. This results in corruption of the breed rather than its expansion.

Upgrading has lots of positives, a few negatives, and has several facets that make it an interesting biological phenomenon. As a backdrop to the issue of upgrading it is important to remember the essential character and utility of breeds. Breeds are useful because they are consistent, predictable genetic packages. It is the predictability that is the key to the value of breeds, for without predictability it is impossible to match breed to place, purpose, and system. Anything that conserves the aspect of breeds as consistent genetic packages tends to be helpful, and likewise anything that detracts from the genetic package is detrimental. Upgrading must always be evaluated in light of this concept.

Anything that conserves breeds as consistent genetic packages is helpful.

The key to understanding upgraded animals is that each of them does indeed have at least the potential for genetic material that is not in the original pure breed. The amount and significance of the outside genetic material are both important. Grades usually begin to very closely resemble the pure breed at levels of 3/4 to 7/8 purebred influence. At these levels, however, they still include a good deal of genetic material that is not from the breed in question. It is important to understand that all of this discussion concerns averages, and individual animals could easily be found that are either a lot more or a lot less "pure" or "pure looking". What this means as a practical issue is that selection towards breed type is especially important in upgrading programs. Upgrading only succeeds if the goal is to add to the pure breed. Upgrading does not succeed if the goal is to change that pure breed into something else.

At higher levels of grade, which certainly include 31/32 or anything higher, the influence of outside genetic material is minimal, and the animals perform and reproduce like purebreds. These upgraded animals may have a slight advantage in overall vigor, and in fact do offer breeds a small breath of fresh air (or fresh genetics) that can be a great boon to some very rare and potentially inbred breeds. Upgraded animals may add lost productivity, vitality, and vigor to these breeds, while also boosting numbers. If upgrading is wisely managed, the upgraded animals pose minimal threat that any of the genetic uniqueness of the breed will be lost. This reflects the basic truth that breeds

Breeds are valuable because they are consistent and predictable.

To contribute positively to pure breeds, such as these Poitou donkeys, it is essential that grades closely resemble purebred animals. It is essential to select for breed type in upgrading programs. Photo by Debbie Hamilton.

are valuable because they are consistent and predictable. Any breeding practice that does not threaten the consistency and predictability of a breed does not threaten its status as a breed, and such breeding practices certainly may include wisely conducted upgrading programs.

The 7/8 grades are generally not sufficiently purebred to truly function as genetic members of the breed, while the 31/32 grades are. This leaves in question the 15/16 animals, which are about 94% purebred. Breed associations that allow upgrading do indeed differ on whether 15/16 animals are sufficiently purebred or not, and the answer to this question has some legitimate leeway when different breeds are considered. These 15/16 animals are generally "purebred enough" to be considered legitimate breed members. Acceptance of males and females is usually at different grade levels because of the potentially broader effect of males on the breed. Some breed associations accept females as "pure" at 7/8, which may be a bit low for inclusion in the pure breed. As a practical issue, though, these females (and it is only females accepted at this level) are mated back to purebred sires for the next generation, resulting in 15/16 offspring (both male and female) that are included in the breed. Some breeds already have reasonable levels of genetic diversity, and therefore have less need for the boost of the small amount of introduced genes that come from upgrading. In those breeds it may be wiser to proceed to higher levels of grade before considering the graded animals as purebred, and indeed some populous, productive breeds probably have little if anything to gain from upgrading.

It is important to remember, though, that if a very high level of grade is required (63/64 for example) then the potential genetic benefits of upgrading will be largely lost to the breed and some very tightly constrained old breeds may need more of a boost than is afforded by introducing only very high grades. There is no single magic cutoff point for the level of grade needed for breed conservation as well as breed vitality. Each situation for each breed must be considered as an individual case, and in no case should upgraded animals be allowed to numerically swamp the older, established purebred stock. Herdbook regulations may be needed to assure that this does not happen. One strategy to accomplish this is to assure that all upgraded animals that are presented as candidates for inclusion into the purebred herdbook pass an inspection and also have a baseline of production characteristics that is seen to contribute to the breed's strength.

Origins of most breeds are mostly less than two hundred years old.

Breed purity and upgrading have enormous political overtones in most breed circles. Breed purity is assumed by many breeders to be absolute, inviolate, and ancient. The truth is that the origins of most breeds reach back less than two hundred years, while most herd books are even younger. Breed formation was initially a response to a desire for predictable animals of a given type. Most breed origins were fairly broad, so genetic viability was assured. As breed registry books were closed and matings occurred only within the narrowly defined breed, the genetic character of breeds changed from fluid and inclusive to very restricted and closed. Some breeds are now suffering varying degrees of inbreeding depression from being locked down with matings occurring only within the purebred population. Continuing to insist on absolute breed purity may well lead

English Longhorns have a long history as the first standardized cattle breed, hailing from the mid-1700s. Photo by D. P. Sponenberg.

to the demise of breeds as they lose their vitality, production, and agricultural relevance. These changes, while initially subtle, result in deviation from original breed type and utility. This has happened in some chicken (Java) and heritage turkey varieties, though breeders are now diligent to reverse that trend.

Dairy goat breeding illustrates some of these issues. Most dairy goat breed associations in the USA register both purebreds and upgrades in separate sections of their herdbooks. The purebred section registers offspring from matings between only purebred members of the registered breed. Upgrades are registered in a separate section and are designated "American." So, a Nubian (or Alpine, or Toggenburg) is a purebred Nubian (or other breed), while an American Nubian is an upgrade. Breed politics are such that upgraded goats have much less demand as breeding animals when compared to comparable purebreds. American goats therefore have lower prices in most markets. Many American upgraded goats, though, outperform their purebred counterparts, so that commercial producers sometimes prefer the American counterpart to the purebred. In this situation the safeguarding of the breed resource as a closed genetic pool has decreased its utility as a viable commercial entity – which is counter to the original aims of breed development, selection, and use! More rational breed maintenance schemes might involve the inclusion of high grades into the purebred herd book, especially if conformation and production levels exceed the breed average. This would assure the addition of only top-quality animals, would still let the breed "breathe" genetically, and would assure that the upgraded goats do not swamp the breed.

Purebreeding and purebreds are vital, so the issue of upgrading does need to be evaluated for its place in breed maintenance, management, and conservation. If carefully managed and operated, upgrading does not threaten the status of any breed as a genetic resource. This is counter to the politics of many breed associations, so upgrading is likely to remain rare. With many international breeds, associations must fit their policies and procedures into those accepted in other countries, and generally the most restrictive country sets the policy on this issue. All breed associations will eventually need to address the issue of upgrading versus absolute breed closure.

Carefully managed upgrading does not threaten the status of breeds as genetic resources.

The downside of completely closed populations is gradually making its power felt as slow erosion of vigor in breeds such as the Thoroughbred horse, Holstein cattle, and many dog breeds. Such declines, if they become widespread, may give breeders cause to evaluate what a breed is, why it is valuable, and how best to manage its genetic status.

Due to the complicated character of breed identity, it is nearly impossible to make general recommendations on specific upgrading programs. Each breed has a unique background, history, and present status both genetically and culturally. At one extreme are breeds such as the Thoroughbred horse, with centuries of purebreeding and a great deal of mystique surrounding its identity and breeding. As a result, any inclusion of outside breeding must be done carefully and in light of the rich history of the breed. One option might be to reinvigorate the breed by a few allowed outcrosses to the Akhal Teke, which is an historically appropriate choice due to its role in the original foundation of the Thoroughbred. It likely makes more biological sense to upgrade some animals to Thoroughbred from a Thoroughbred x Akhal Teke base in an upgrading plan, rather than including early outcrosses as purebreds and thereby run the risk of a sudden change in the genetic makeup of the breed. By this strategy the breed package remains intact and loses no predictability. In contrast, a widely divergent cross such as to a Brabant Belgian has little to offer the breed genetically or culturally.

In contrast, relatively recently defined landraces, such as the Pineywoods cattle or Navajo-Churro sheep, have to consider upgrading very carefully. These breeds, though centuries old, retain diversity and face the challenge of breed definition even more than they face any challenge of viability. For these breeds upgrading, at this point in time, offers little if anything.

Upgrading, if carefully managed and monitored, can assure that pure breeds remain viable and vital. Upgrading can also assure that breeds retain their status as genetic resources. Any upgrading program must be carefully monitored to

The Kensing line of Spanish goats combines productivity with range adaptation. Photo by D. P. Sponenberg.

assure that appropriate levels of grade are achieved before inclusion of animals into the purebred breed. One strategy that assures this is the registration or recording of all the generations of grade so that the association is confident of the genetic status of each animal. A final hurdle that might be used at the generation included in the main purebred herd book is to inspect those animals, either by photograph or physically, to assure congruence with breed type.

At the same time that breeds protect themselves from too radical an inclusion of grades and outside influences, it is also important for breeds to recognize the value of grades to the viability and production of the breed. The value of grades can be quickly negated by the political view in many breed associations where upgraded animals are considered second-class citizens. For a breed to fully reap the biological benefits of upgrading (numerical expansion, increased market for breeding males, some gain in genetic vigor), the upgraded animals must at some level of grade indeed be considered as full members of the breed. That is where breed politics get into the picture, and breed politics frequently do not have an answer in biology.

The controversies of whether or not to allow upgrading do not need to generate huge amounts of passion. Some breeds have little to gain from upgrading, but few have much, if anything, to lose. The final determination of the appropriateness of upgrading for any breed will be decided by tricky issues such as breed politics and reciprocity between national herd books, which is absolutely essential for international breeds. Upgrading in some form does, however, make good genetic sense for nearly every breed, but only if upgraded animals are eventually included as full members of the breed with all benefits, privileges, and respect that are afforded other breed members.

An example of breed association rules prohibiting upgrading and therefore undermining effective reciprocity of an international breed comes from Wensleydale and Teeswater sheep. These two rare luster longwool sheep breeds are English, and only recently has semen become available for use in the USA. Because importation of live sheep from Britain is not allowed by USA regulatory authorities, no Wensleydale or Teeswater ewes and rams can be imported, only semen. Wensleydale and Teeswater breed associations in the USA insist on the use of foundation ewes that are either Lincoln, Cotswold, or Leicester Longwool, all of which are rare breeds globally. The short generation interval of sheep allows breeders to rapidly grade up to nearly purebred Wensleydale and Teeswater sheep, but these upgraded sheep are not recognized as pure in the United Kingdom, the breeds' country of origin. As a result the whole procedure has tended to extract ewes from the limited and valuable ewes of the three rare longwool breeds already in the USA, and has not effectively contributed to conservation of the two rare imported breeds because the upgraded sheep are not recognized in the United Kingdom. They therefore cannot be reciprocal between the USA and the UK; they will likely never be fully recognized as international members of those two breeds.

Upgrading and Bloodlines

Bloodlines within a breed are important, and have been discussed in earlier chapters. Maintaining these as genetically healthy resources for the breed is important. Similar to the situation of a breed as a whole, bloodlines within a breed also gain important and necessary benefits by upgrading using other lines within the same breed. The issues of upgrading to a pure breed from crossbreds are slightly different than those concerned with grading within a single pure breed to a specific bloodline within the breed. Having distinct bloodlines within a breed helps breeders to manage the genetic material of the breed so that relatively isolated pockets within the breed maintain some genetic distinctiveness. This can help to maintain viability in the breed, but only if these bloodlines remain viable and vital. This is accomplished by the judicious use of upgrading with other bloodlines, which starts by linecrossing.

The level of genetic distinction between bloodlines is much lower than that between breeds, because bloodlines within a breed are indeed all of the same breed. The level of pure strain breeding of a candidate animal to be considered as being of a strain or bloodline is therefore not as high as that required for inclusion of a true upgrade into a breed. This is, of course, only true if the grading animals are of the same parent breed. A purebred animal that is only 7/8 of a specific bloodline is for most purposes essentially of the bloodline in question.

The Cotswold is a globally rare luster longwool breed that has had ewes diverted to upgrading for Wensleydale or Teeswater breeding programs in the USA. Photo by SVF Foundation.

106   Maintaining Breeds

The inclusion of 7/8 or 15/16 animals into old established bloodlines can meet some political resistance, but is essential if older bloodlines are to remain viable and contributing members of the breed. To lock them down genetically is to assure their eventual demise, and they offer too much benefit to breeds to allow that to happen.

Upgrading to Rescue Endangered Bloodlines

One very useful application of upgrading is the rescue of bloodlines that have become endangered. Especially in landraces it is common to encounter old family strains that have been reduced to a handful of older females and no males. In this situation it is all too common for breeders to then use a sequence of outside males, the result being a steady dilution of the original genetic material of the strain. The first crop of offspring is 1/2 the original strain, the next generation is only 1/4, then 1/8, until very little of the old strain genetic material remains.

An alternative that concentrates rather than dilutes the original strain is to use an outside male on the original females, then save at least one and preferably two or three male offspring. These are then used, at the youngest possible age, back on the original females. The resulting offspring are then 3/4 of the original strain, and ideally the tactic of using new young males as soon as possible is repeated to yield a crop of offspring that are 7/8 the original strain.

Success depends on the number of the original females, their kinships to one another, their ages, and their longevity. In many cases it is possible to generate a population that is largely the original genetic material of the old strain. This can be useful for overall breed diversity and genetic health. While the general outline of a strain rescue is presented here, each case is unique and specific plans must be tailored to each specific situation.

For example, one Tennessee Myotonic goat bloodline comes from Sarah Eggerton of Texas. These goats were down to three females, and lightning killed

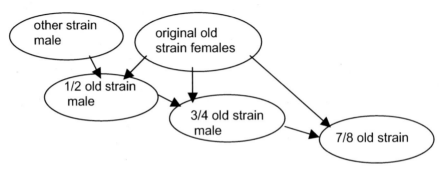

A bloodline can be rescued by using a strategy that wisely uses older females of the line as well as their sons.

The Agricola strain of Pineywoods cattle is being expanded through an ugrading program. Photo by D. P. Sponenberg.

the sole remaining buck. A son of one of the does was produced by another bloodline of buck, resulting in a buck that was 1/2 Eggerton. While this initial buck was growing, a second buck of yet another line produced some 1/2 Eggerton kids. These bucks were used back on the Eggerton does, resulting in bucks and does that were 3/4 Eggerton breeding. One of these was used back on the sole surviving doe, providing a buck that was 7/8 Eggerton, and he was able to be used on some of the daughters of the other does, providing for a group of animals that had between 1/2 and 3/4 the influence of the old Eggerton line. With an earlier start and more initial females this rescue could have generated a group with even higher proportion of Eggerton breeding, but considerations of eventual inbreeding limited the final percent Eggerton breeding that was possible.

Upgrading within a breed can rescue endangered bloodlines.

A similar bloodline rescue is ongoing with the Agricola line of Pineywoods cattle assembled by Luther Schell and Bo Howard. Fortunately multiple cows are available, and with these higher numbers it is possible to get a higher population of 7/8 Agricola cattle than was possible with Eggerton goats.

Adaptation

Adaptation is the ability of animals to survive and produce in compromised environments. For many breeds this is an essential component of their history and their usefulness, and Dr. Stefan Adalsteinsson refers to many breeds as having "the genetic heritage of survival" to capture this important concept. Maintaining adaptation is sometimes a very tall order. This is especially true for some breeds as they move from a landrace situation to one based on the standardized breed idea. When breeds become popular they tend to move from being a local resource to more of a national resource. This is the usual situation because of increasing demand for the breed, an increasing price, and with these an increasing incentive to concentrate on production and eye appeal rather than on survivability in compromised environments.

For selection to favor adaptation, the environment must be managed to demonstrate differences in levels of adaptation. This really translates into allowing some of the animals to demonstrate a lack of adaptation, and to cull on this basis. Such a system allows some animals to fail. This can be frustrating or distasteful to breeders who are more comfortable providing a supportive environment that favors maximal opportunity for animals to be productive. This situation also raises a moral dilemma as to how "hands off" management should be while allowing for adaptation to express itself. When treatments and resources exist to counter nutritional or infectious challenges, it is questionable to forego their use, even though using these inputs may well mask differences in adaptation.

Adaptation is one slot for animals to put resources; production is another.

For example, the Tennessee Myotonic (or Fainting) goat is reputed to have some resistance to gastrointestinal parasites. This is an important trait in the breed, which will only gain in importance in the future. To assure that parasite resistance remains present in the breed, breeders must not deworm every goat at three-week intervals, but instead must allow the worms to build up, infect the

Tennessee Myotonic (Fainting) goats have some resistance to parasites, but must continue to be selected for this useful trait. Photo by D. P. Sponenberg.

Some lines of Texas Longhorns have been selected away from this original rugged range type of the original breed. Photo by D. P. Sponenberg.

susceptible goats, and then select as breeding stock those that resist the effects of the worms. In most situations the result is that some goats will be wormy and relatively poor-doers to demonstrate which ones are worm-resistant and productive. This situation presents both economic and ethical problems. Fortunately, the development of strategic deworming tactics based on level of anemia (and thereby worm load) has allowed selection for parasite resistance to proceed without allowing losses from parasitism to occur. The key here is that breeders need to be diligent to act on the information of relative anemia so that replacement stock is kept from the most resistant animals. Breeders should be eager to track the performance of their goats with respect to parasite resistance, so that future generations can benefit from decreased reliance on deworming chemicals.

The issue of selection for adaptation is nowhere certain in the minds of breeders and academics. To what extent any breed can have maximum resistance and maximum production is very much an undecided question. Some evidence points to the concept that animals have limited metabolic resources at their disposal, and so they must "choose" where to put those. Adaptation is one slot for these resources, production is another. While it may not be as simple as that, this approach does seem to have some merit. Breeders may well have to decide on the relative importance of adaptation and production for their own systems and their own breeds. A low-input, adapted breed will likely never equal the production levels of an industrial strain in a high-input system. What is lost in productivity is compensated for by lower inputs such as housing, dewormers, refined rations, antibiotic support, and veterinary interventions during birthing or at other times.

Some breeds have undergone very rapid transformation from adapted, rugged animals to more smooth and productive ones. The Texas Longhorn, especially the show ring lines, has certainly done this, and breeders of other breeds will be tempted take their breeds down this same road. As the type changes, and the selection environment goes from compromised deprivation to a life of ease and plenty, so too go the genetic resources forged by that environmental change. These issues have no easy answers, yet are essential to consider in order to arrive at some sort of meaningful philosophy of genetic resource conservation.

Some breeders have managed to select for adaptation as well as for production, and have produced some interesting genetic resources in the process. Over a thirty- to forty-year span, Robert Kensing of Menard, Texas took the original local Spanish goats (60-pound females) and selected them from within the breed (no crossing) to produce females that were 150 pounds at maturity. At that size he noted less adaptation to the rugged west Texas conditions, and so he relaxed selection for size and ended up with 125-pound females that were fertile and productive as well as adapted and able to use the rough browse of the environment to raise kids without supplementation of other feedstuffs. It is possible to have adaptation and production in some combination, but it takes very wise and dedicated breeders to assure that the drive for production does not leave adaptation out of the equation.

Long-term Genetic Management of Breeds

Managing breeds for long-term survival is a difficult balancing act. Breeds must retain their predictability, based on relative genetic uniformity, in order to be useful. In addition, they must retain levels of genetic variability that provide for adaptation and viability. Close attention must be paid to bottlenecks, founder effect, and the relative contribution of individuals to the next generation of animals. To avoid considering these is to run the risk of a narrowing of genetic variation that can threaten breed viability. Another major issue for breeds is upgrading and whether or not to allow it in the management of the breed. Upgrading must be managed in order to not change the breed by introducing too much variability, although it does offer breeds an opportunity to remain genetically viable while protecting breed genetic identity.

5. External Factors Affecting Breeds

Forces that are external to breeds and their breeders can have important influences on breeds and breed survival. These influences can be either positive or negative. External influences must be recognized and managed for breeds to succeed. If breeds cannot serve in an environment dictated by external factors, then they quickly become trivial and only of interest to a handful of dedicated breeders keeping the breed as a hobby. While the hobby approach can and does succeed in saving breeds from extinction, conservation of breeds serving some demand imposed by the outside is a much more effective long-term strategy for breed survival. Breeds that are useful have a much better chance of long-term survival than do those that find no appropriate practical use – provided that narrow single-purpose uses are not the only yardstick used in making the determination of the usefulness of a breed!

Market Demand

Market demand for breeds can have a huge impact on breed numbers, but must be carefully weighed against other motivations for stewards of rare breeds, and these vary dramatically. At one far end of the continuum are people who are motivated by a sincere concern for biodiversity of a particular breed, as well as safeguarding and expanding a genetic resource to fit into a production niche. At the other end are those who are cynical and self-serving, and have motives that stem from a desire to cash in on the demand for rarity or to enhance their own sense of importance. In between these two extremes lie most breeders, who are interested in breed survival but also have concern for a reasonable and justified economic payback from the breed. Secure market demand can bring security to a breed by providing positive outcomes for the entire spectrum of breeders.

An essential aspect of breed survival is demand for the breed. It is fortunate for the long-term survival of many breeds that rarity itself is appealing, so that in the present era the truly rare breeds are unlikely to undergo complete extinction. Nearly every breed can indeed survive on the basis of interested conservation breeders motivated by concern for avoiding breed extinction. The downside of this approach is that such breeds can quickly become viewed as oddities with no real purpose. Most rare breeds deserve better because they are useful and productive in the appropriate habitat and should ideally remain part of a functioning agricultural landscape.

For breeds to be sustainable it is ideal if the demand for the breed comes directly from its products or services. Where breed products and services are desired, demand assures continuity of the breed. Healthy commodity-based demand assures that breeders will be monetarily rewarded for their efforts in producing the breed, giving breeds a secure niche in agriculture.

Breed demand can fall into several categories. A very healthy situation for a breed would be for it to fit market channels that already exist. Products, whether meat, fiber, eggs, or milk (or services such as draft, grassland management, pest control, and wildfire fuel abatement) might currently fit what is desired. Such breeds with "ready-made" markets are indeed the rarest of the rare, largely because if they already fit mainstream market demand, then they would have never suffered rarity in the first place. Some few such breeds do indeed exist, however, and their reasons for rarity usually rest in some factor other than production. For breeds such as the Red Poll, for example, the fit is already there with mainstream agricultural demand. The challenge is to make that connection more obvious to candidate breeders in order to enhance demand for this productive breed.

Useful breeds have a long-term future.

Rarity in and of itself does bring some demand, but a market based solely on rarity has potential problems. Some breeders have tried to assure or prolong breed rarity, usually by overly restrictive registry procedures, in order to

Four-horned Navajo-Churro sheep are much more than an oddity; they also have important ceremonial purposes in Navajo culture. Photo by Tanya Charter.

preserve the market advantage they have by virtue of being already involved in the breed. Self-serving breeders have also occasionally become breeders of a rare breed, but then quickly get out of the breed as numbers increase and prices fall from the highs induced by the novelty or exotic market. When breed rarity becomes an end in itself, it invites a departure from linking breed demand to any commodity or production concerns. At that point, the market is likely to serve conservation poorly, if at all.

Products that many rare breeds produce fall outside of the norm for agricultural commodities. Examples are the many rare sheep breeds that produce smaller lamb carcasses than the relatively large ones desired by the large meat packers that control most of the mainstream commerce in lamb meat in the USA. In such situations it can be necessary to develop and serve alternative markets that create increased demand for specialized products. These can easily include breed-specific or breed-labeled products where the uniqueness of the product becomes an asset instead of a liability. A recent phenomenon has changed what have been minor specialty markets (small lambs, for example) into more mainstream demand. To the extent that the market now readily accepts more variable products than a few decades ago, more and different breeds find a production niche more easily. Some early success with acorn-fattening of very fat hogs for specialty ham production further illustrates how the match of the right breed, right process, and right market can dramatically change the fate of a breed.

The amazing turn-around in the fate of standard turkey varieties testifies to

Red Poll cattle have already been proven to be tops in calf production, and have a ready-made commercial niche. Photo by Nathan B. Melson.

Matching heritage turkeys to a targeted market has sparked resurgence in turkey numbers and secured these birds a bright future. Pictured is a handsome pair of Chocolate turkeys. Photo by D. P. Sponenberg.

the potential benefits of product-specific promotion to foster breed conservation and expansion. Populations of nearly all varieties of heritage and standard turkeys had collapsed by the late 1990s. A targeted program by ALBC and Slow Food USA then increased the demand for these birds as holiday fare, and the numbers of breeder birds furnishing poults for growers serving these markets has steadily increased year by year to provide these historic varieties with a more secure place in the agricultural landscape.

Promotional programs for rare breed products can sometimes be very effective across several breeds, such as the "Save the Sheep" contest run by *Spinoff* magazine. The effort involved inviting participants to handcraft items from fibers produced by rare breeds. This generally involved wool, but was open to other fibers such as cashmere from rare breeds of goats. The effort specifically targeted rare breeds, and provided education about them.

The result of the "Save the Sheep" program was an increased awareness on the part of handspinners and others as to the wonderfully wide array of fibers produced by rare breeds of livestock. The program increased demand for fibers produced by pure breeds, which boosted income to producers. In addition, a handsome book was produced which detailed the whole subject of breed uniqueness and breed rarity. This effort successfully targeted an audience that was generally uninformed about issues of rare breed conservation, but is uniquely and powerfully situated to help conservation efforts by linking fiber purchases to rare breed conservation programs. The book, *Handspun Treasures from Rare Breed Wools,* continues to educate handspinners and others about breed-specific products.

Crossbreeding

Several rare breeds are rare for the peculiar reason that their crossbred offspring have high value. For some of these breeds it makes more economic sense to use purebreds for crossbreeding rather than for the production of purebred offspring. The result of such a market imbalance is that purebred recruitment (the production of purebred daughters and sons) can become dangerously low as high numbers of purebred dams are crossed rather than bred pure. The next generation of purebreds is simply not being produced, and breed numbers dwindle.

Excellence in crossbreeding has endangered several breeds. Spanish cattle in the Americas (loosely termed "Criollos") are a good example, and include the Florida Cracker, Pineywoods, and Texas Longhorn cattle in the USA. These cattle survive and produce well in compromised environments, and are genetically distinct from other cattle. Because of this distinctiveness they produce vigorous and productive crossbred offspring with both Brahman cattle as well as Northern European cattle. The excellent survival and production ability of half Criollo calves has led to purebred Criollo cows being put to bulls of different breeds, with retention of the crossbred heifers. What has been lacking is enough pure breeding to assure that the pure Criollos are available for crossbreeding in the future. Several countries have come to mourn the loss of the purebred Criollos as overall cattle productivity has fallen when the Criollo influence has

This half Angus, half Florida Cracker calf demonstrates the boost that is achieved from crossbreeding cows of several Criollo breeds. Photo by Tim Olson.

dropped below 25%. This pattern has been repeated with local types of criollo cattle in most countries of Latin America.

Other examples of crossbreeding leading to purebred decline include several horse breeds. These include some warmbloods, such as the Cleveland Bay and Irish Draught, as well as a number of pony breeds such as the Connemara. The heavier utility type of these breeds can produce wonderful sport horses when put to Thoroughbred mates. In some cases the resulting crossbreds have been included in the herd book, with the result that the older, heavier, utility type is mixed in with the lighter, athletic, modern sport type. As breeders select breeding stock they tend to favor the modern athletic type without regard to its mixed pedigree, and by this process the original type becomes rare or lost. This ignores the fact that the modern sport horse is a hybrid (and a good one, at that) springing from the traditional utility type. Very few horse breeds have been able to counter this trend. One strategy to avoid losing purebred type is the use of a "part bred" herdbook which lists the crossbreds, recognizes the parent breed's influence on them, but does not allow them to replace the traditional type as breeding stock. Such a strategy very effectively recognizes the excellence of these crossbreds (athletic performance) while protecting the traits that the pure breed has (reproducing athletic excellence through crossbreeding).

Excellence in crossbreeding is especially perplexing as a cause for breed rarity, because it is the breed's very excellence that is endangering it. In a few situations breeders have been able to capitalize on crossbreeding excellence while conserving the pure breed. Blanco Orejinegro cattle, a Criollo breed of Colombia, are prized for their resistance to certain parasites. Breeders of these cattle are using purebred bulls on commercial dairy cattle to impart this resistance. This works better for the breed than does using Holstein semen on the Blanco Orejinegro cows. The path these breeders have followed has saved the breed from extinction so that the breed can serve future generations with its useful characteristics.

Regulations

In some situations external regulations, usually governmental, can have profound effects on rare breed conservation. Recent moves in the European Union and several European countries to insist upon scrapie-resistant genotypes in breeding rams have certainly changed breeding choices in several breeds. To some extent this trend is being mimicked by breed associations in the USA, as well as by some individual states. This can rapidly change breeds by limiting choices of breeding stock. When a ram fails to be used solely because of his scrapie-resistance genotype it is not only the single gene that fails to be passed along – it is his

Bull licensing laws in the Netherlands nearly caused the extinction of the Dutch Belted breed in its homeland. Photo by D. P. Sponenberg.

entire genome that has been culled.

In contrast, the Australian approach has been to insist that imported sheep be of susceptible genotypes, but free of scrapie. This approach assures that regardless of the biology of resistance (long incubation versus true freedom from the disease agent) Australia is very unlikely to be importing any problems along with select carrier sheep.

Regulations profoundly affect rare breed conservation.

The USA is somewhat different than most other countries in having very little governmental oversight of animal breeding. In several European countries, for example, only licensed, registered bulls and stallions are allowed to reproduce. This can and does have a profound impact on the status and type of breeds. In some instances this is good because officials can have a traditional and conservative eye for type, such as occurs with the Noriker horse of Austria. In other cases, though, contemporary fads can drastically change a breed and due to the government's participation little can be done to counter this, such as occurs with Warmblood horse breeding throughout most of Europe.

Conservation can thrive under many different specific rules, as long as the philosophy is sound.

In any situation, no matter how regulated, a key issue is whether effective conservation is taking place by virtue of the practices that are in place in purebred selection programs. Conservation can thrive under a variety of specific rules, as long as the philosophy of guarding breed type is sound.

Imports

Rare breed imports present a complicated array of issues for breed conservation. Some importations of rare breeds are a benefit to conservation while others are a threat to conservation both here and in their country of origin. Nowadays, imports generally come into this country with good documentation, at great expense, and are select stock. This gives the imports an instant advantage in credibility and publicity over numerous American breeds that are in peril, and most especially those that are raised in extensive or traditional situations because the breeders in those systems have rarely participated in breed organization or promotion. While some imported breeds are likely to be involved in constructive conservation endeavors, others are likely to detract from serious livestock breed conservation in the USA and in other countries.

Imported animals present important issues for rare breed conservation.

The effects of imported breeds vary, and any summary is bound to be inaccurate by not reflecting a host of finer details. With that disclaimer, it is possible to put imported breeds into four general categories:

1. Imports that contribute substantially to conservation efforts.
2. Imports that enhance American bloodlines.
3. Imports that hamper conservation in their country of origin.
4. Imports that endanger American bloodlines or breeds.

Imports That Contribute Substantially to Conservation Efforts

Imports that contribute to conservation are success stories, and are therefore the easiest and most pleasant to discuss, largely because this is a pleasant topic. Breeds that fit into this category include those that are endangered in their homelands, usually by political or geographic threats. The Caspian horse, for example, is endangered in its Iranian homeland. The danger was once from political upheaval. Today it is more from lack of a targeted conservation program in Iran. Imports from Iran are more likely to be used for conservation in their new country (primarily in the UK, but also in the USA) than in Iran. The goal with this breed and others in a similar situation should be to foster meaningful conservation in their country of origin. In the absence of such programs, though, conservation is best served by the representatives of the breed that have been exported to other countries.

Other importations that contribute substantially to conservation are somewhat surprising. One is the Shetland goose, a small and distinctive goose breed imported by Nancy Kohlberg of Cabbage Hill Farm in New York. The geese in

Shetland geese have become a secure genetic resource in their new American homeland. Photo by D. P. Sponenberg.

the American population have increased their numbers beyond those in Britain, and in the process have significantly contributed to the array of divergent goose breeds available in the USA. Through the efforts of Cabbage Hill Farm in New York and Holderread's Waterfowl Conservation Center in Oregon, these geese have become more globally secure both genetically and numerically. While this is indeed a success story, all of the American Shetlands descend from only one foundation population, and conservation genetics principles indicate that a more secure base would have included a few birds from other populations. This may only be a theoretical issue, though, as the present birds are vigorous and productive.

Poitou asses are another imported rare breed success story, largely through the efforts of Debbie Hamilton of Vermont. Her assurance that the breed in the USA is reciprocal with the breed in France is the result of long hard work. This specific effort has made it possible for the small American representation of this breed to have an important impact on the breed's survival in its homeland as well as here in the USA. A genetically significant portion of the breed is in the USA, and Debbie's diligence assures that the international breed can benefit from it.

Some importations could not have been predicted to lead to significant contributions to global breed conservation, but have ended up doing just that. Leicester Longwool sheep were imported by the Colonial Williamsburg Foundation as an appropriate breed for their historical interpretation program. With thriving and at least somewhat commercially relevant populations of this breed in the UK,

The elite French mule-breeding Poitou ass has been strengthened by the reciprocity of French and American breeders. Photo by Don Schrider.

Australia, and New Zealand it was difficult to foresee that the USA's contribution would be significant for this breed. Fifteen years after the importation, the numbers of this breed in the USA have risen while those in Australia and New Zealand have plummeted. The US population, based on Australian, New Zealand, and English breeding, has therefore emerged as an important genetic reservoir of the various international strands for this old and unique breed.

A common thread for most of these, as well as for other, importation success stories is that the imported animals of several of these breeds are more likely to contribute to conservation following importation to the USA than they were to make contributions in their home country. This is unfortunate, as breeds are always best conserved in the region from which they originated. For breed conservation to be successful with these imported animals it is essential that a broad genetic base be imported. In addition it is important to assure that the imported animals are managed for maximum genetic viability with good representation of all founders. A casual, nondeliberate breeding program is much easier in the short term, but is very unlikely to succeed in accomplishing long-term conservation goals.

Imports That Enhance American Bloodlines

Imported additions to existing breed populations in the USA offer potential conservation benefits. These must be used carefully, or they risk diluting the uniqueness of American populations while at the same time invigorating and broadening them. For some breeds, such as the Clydesdale and Shire horses, such importations to the USA are coupled with exportations to other countries with populations of these breeds. The reciprocal transfers keep these breeds functioning as important and vital international breeds. The recent importation of semen from New Zealand and England to bolster Leicester Longwool sheep in this country is an example of enhancing American bloodlines through importation. These breeds are indeed international rarities, with several countries contributing to their conservation. Occasionally exchanging genetic material among the countries assures that each of these breeds remains a single breed rather than fragmenting into multiple mutually exclusive national populations.

Imports to enhance bloodlines should be occurring in all directions so that the USA does not become a net importer with no associated exportation. With no genetic material going out, the result is that our populations become a dead end for influences from other countries but with no influence in the reciprocal direction. This can easily negate an important role of American populations as reservoirs of genetic variability. Import and export restrictions frequently dictate the ease with which these international breeds become reciprocal between nations. The goal should always be easy reciprocity to assure that the international breed remains a single gene pool.

Imports That Hamper Conservation in the Country of Origin

Imports damage the breed in the country of origin when irreplaceable bloodlines or animals are removed from conservation programs in their country of

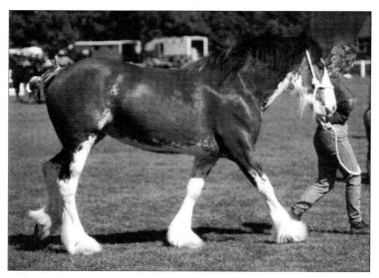

The Clydesdale horse is a very good example of a truly international breed, with genetic exchanges among all the countries breeding this handsome draft breed. Photo by D. P. Sponenberg.

origin. Fortunately examples of these are relatively few. In the last few decades the export of many Galloway and Highland cattle from the United Kingdom into Germany has resulted in some drain on those breeds. These imports were less likely to remain documented or to contribute to purebred breeding programs, and therefore many were lost to the breed.

Imports damage their breed in its homeland when irreplaceable bloodlines or animals are removed from conservation.

A similar phenomenon can and does occur in horse breeds when high-performing and select breeding stock is exported. This can result in a slow and steady drain of the top of the breed from the country of origin. The breed in its country of origin thereby suffers an erosion of quality and potential, and can become relegated to second-class status among countries holding the breed. This is never in the best interests of breed conservation, although for a few breeds such exports are the only hope that some animals will be used for breed conservation as they are unlikely to be used in the home country for purebred breeding.

Imports That Endanger American Bloodlines and Breeds

Imports can endanger uniquely American breeds as well as uniquely American bloodlines within American breeds. The negative effect of imports is usually by direct competition for space on farms. The competition disadvantages local genetic resources, with the common result that the imports supplant either breeds or bloodlines that are impossible to replace. This group of imports therefore demands close attention.

Several international beef breeds can provide examples of this category. Devon cattle are a very good example. Some countries, notably excluding the USA, allow at least some crossbreds or grades to be included as purebreds. Thus Devon cattle in this country have largely avoided inclusion of Salers breeding, and as such serve as a reservoir of a more original genetic resource for this internationally distributed breed than is true of many other countries.

A new and paradoxical threat has emerged for Devon cattle in the United States. As the Devon has become valued for its excellence in grass-fed beef production, demand for the breed has grown. Several breeders have promoted the use of semen from New Zealand Devon bulls in an effort to enhance grass-fed potential in the breed. The result has been a steady decline in purely American bloodlines, and the American portion of the breed is indeed in danger of disappearing through what is essentially upgrading with New Zealand sires. This is ironic as the full potential of the American lines could well be lost before they are even documented.

As Devon cattle become increasingly international the American bloodlines run a real risk of replacement by newly imported genetic stocks from New Zealand. Photo by D. P. Sponenberg.

While no recent imports have occurred with Jacob, Karakul, or Tunis sheep, the same potential threat exists for each of these breeds. The American type of each of these breeds is distinctive, and in some instances closer to the original than what could now be imported. If breeding stock for these breeds were to be imported, the result would be a decline in the uniqueness of the American bloodlines. Another example of a breed that could be damaged by importation is the Dutch Belted dairy cattle breed, because America has long served as the major reserve of genetic material. The Dutch population is at this point very distinct, largely as a consequence of bull licensing laws, which for decades discouraged the use of belted bulls.

In some situations, the threat to American bloodlines, such as with White Park cattle, is more subtle. Most American bloodlines of this breed are the result of importation in the mid- 1900s, and occurred before a genetic bottleneck in Britain. American bloodlines were then augmented by a recent importation of semen, and fully pre-bottleneck animals are now rare. Pre-bottleneck cattle remain valuable because they have a potentially great contribution to the long-term survival of this ancient and pivotal cattle breed.

In another example, a Sorraia stallion has recently been imported from Portugal, and some breeders plan to use him on Colonial Spanish mares within the USA. While superficially similar in type, the mares are from a variety of origins and crossing between branches of a family that have been separated for several centuries could well end up displacing some of the unique attributes and genome of the Colonial Spanish horse in the USA. This is not likely to occur to a very damaging degree, but the level of interest in an imported horse is greater than that in locally produced ones. Because of this attraction, imports are likely to displace locally produced animals of similar quality and higher conservation value.

The phenomenon of imported animals having better name recognition and higher cost than local animals can have a surprising result on popularity in some cases. The various branches of the Colonial Spanish horses in North and South America closely resemble one another. The local branches of this family are all rare, with some in peril of extinction. In spite of this, the imported Peruvian Paso, Paso Fino, Mangalarga, and horses of other breeds enjoy greater popularity and command a higher dollar value even though similar North American horses could be had for a fraction of the price of the imported model. This phenomenon does great damage to local breed resources as it focuses attention on the imports, each of which takes up a stall that could be housing a locally produced horse. These imports divert attention away from uniquely American genetic resources.

Some imports do more than displace bloodlines; they instead displace entire rare American breeds. They usually do this by occupying barns, stalls, and pastures that would otherwise be devoted to raising rare breeds already in this country. These imports include some breeds that have become real threats to the maintenance and survival of our unique breed resources. Some of these are relatively common breeds that are recently imported, such as the Boer and Kiko goats whose impact is felt by Spanish and Tennessee Myotonic goats in this country. Dorper hair sheep are displacing Katahdin, St. Croix, and Barbados Blackbelly

St. Croix sheep are threatened by crossbreeding to more recently imported breeds of hair sheep. Photo by D. P. Sponenberg.

sheep. Icelandic sheep tend to displace at least some Navajo-Churro sheep, although this will never be complete because the environments of origin for both the Navajo-Churro and Icelandic sheep are far different and too harsh and demanding for most other breeds. The Icelandic sheep only competes with the Navajo-Churro outside of the original environment for the Navajo-Churro sheep.

In an effort to document the potential of Iberian-based cattle breeds for beef production in the South, the USDA imported the Colombian and Venezuelan Romo Sinuano breed. This is a useful and productive breed, and is being used in breed comparisons in studies in Florida. The local Florida Cracker, a cousin breed to the imported Romo Sinuano, is in dire straits and is not included in these studies. Private endeavors as well as public ones often repeat the illogic of ignoring the local resource, documenting an imported relative, which then displaces the local resource. This generally happens at great expense and effort, but happens nonetheless while similar and local genetic resources plunge toward extinction.

The importation of Wensleydale and Teeswater sheep semen has posed a direct short-term threat to other luster longwool breeds in this country, such as Cotswold, Lincoln, and Leicester Longwool. Ewes of these three breeds are designated by the USA breed associations as the only recognized recipients of this semen to establish a base for upgrading to high-grade Wensleydales and Teeswaters. The irony of this situation is that the breeds from which these ewes come are rare in their own right, and in some cases more rare than the imported resource. These breeds are U.S. representatives of internationally recognized breed populations that contribute to international conservation for all three. The Wensleydales and Teeswaters in the USA, however, will be the result of upgrading and will likely never be recognized as full Wensleydales or Teeswaters in their country of origin due to breed regulations in the United Kingdom. Upgraded sheep will therefore never be able to contribute to international breed numbers or conservation efforts for Wensleydales and Teeswaters. This situation has made one rare breed directly threaten others, with no long-term contribution to global security of any of the five breeds involved.

A further conservation implication in the luster longwools (Leicester Longwool, Lincoln, Cotswold, Wensleydale, Teeswater) is that these breeds do indeed tend to compete directly with one another for space on farms. In the United Kingdom it has been shown that the overall numbers of longwool sheep remain fairly constant, but that within that number the relative numbers of the individual breeds fluctuate. The situation in the USA is similar, so that a space taken by a Wensleydale or Teeswater upgraded sheep is very likely to have been filled by a purebred Cotswold, Lincoln, or Leicester Longwool rather than by a crossbred or non-longwool sheep. This becomes a zero-sum game in many instances, with the import displacing and preventing meaningful conservation of another breed in danger and already established in this country.

In a somewhat similar vein, Shetland sheep have surged in popularity. When first imported to the USA this breed was on both the British Rare Breed Survival Trust and the ALBC rare sheep breed list. It is now a conservation success story, and has become much more secure numerically. A large part of the reason for that is not only the many advantages these sheep have for smallholders, but also the fascination with a breed from a remote corner of the globe. Many sheep enthusiasts keep Shetlands, therefore, while the multicolored and finer-fleeced Romeldale remains an American breed in very real danger of extinction.

Several horse breeds are similarly problematic. Each horse occupies a slot that then becomes unavailable for other horses, and the endangered American breeds must compete with several imported breeds for attention. The imports include many British pony breeds, for which the breed dynamics very much result in America being a one-way street, rather than a key participant in the breed's international survival strategy. The imports are contributing little to conservation of their own breed, and at the same time are also taking up spaces that horses of uniquely American origin could have occupied.

Other imports are more difficult to classify as to net effect on global conservation. Effective conservation of the primitive and feral breeds is problematic outside of their original habitat. Soay sheep, Arapawa goats, and Exmoor ponies belong in this group. It is important to question whether any of these breeds that is so closely associated with a specific distinctive environment can be effectively conserved outside of that environment.

Assessment of Importations

Imports of rare breeds are frequently problematic for conservation. The reason for this is that imports tend to be cared for by dedicated individuals with great enthusiasm and loyalty. These are the very sorts of advocates that are needed for the rare breeds already endangered in this country, and their energy and talents would have greater effect focused on the high priorities that are already here rather than importing new priorities. Exceptions do exist where imports do not hamper conservation priorities and do indeed accomplish significant conservation work. While these few cases are worthy of celebration as valuable conservation successes, it remains a fact that imported breeds have little positive effect for conservation of rare breeds, and many do a great deal of damage. Any program with imported, especially recently or newly imported, breeds needs to be very carefully evaluated as to its consequences both for the breed in question as well as for other endangered breeds.

6. Associations

Breed associations are groups of people with a shared interest in a specific breed. While this may seem obvious, the fine details can become complicated and can splinter rather than unite people. Issues involved in breed associations are vitally important to the maintenance and conservation of pure breeds of livestock. Some of the issues are procedural, others are political, but all must be addressed for breed associations to do their job successfully.

Breed associations are vitally important to the maintenance of pure breeds of livestock.

Many breeds are rare because of political and organizational failure on the part of breed associations. Success for all breed associations has a similar array of characteristics:
- timely communication
- inclusiveness
- frank discussion without rancor
- efficient processing of registrations and other communication details.

Failure in any one of these can lead to a weak and ineffective breed association that does the breed more damage than good. Breeds must not only survive the physical environment, they must also survive the political environment of their own breeders! In a few cases the internal threats to a breed's existence are the most significant ones, and overshadow any biological or market issues challenging the breed.

Understanding the philosophies and characteristics of breed associations can help them to work smoothly. Some associations are driven almost completely by market forces. These include some very large and powerful associations that govern popular and numerous breeds and for which economic forces are predominant. Other breed associations tightly focus their activities on issues of breed maintenance and genetic identity. Still others are most concerned with social interaction among the breeders of the breed.

Many breeds are rare because of political and organizational failure.

People join associations in order to belong to a group of people with shared interests. It is important for all breed associations to reflect on group identity and goals so that the activities of the group can serve the breed effectively. Successful associations, especially for rare breed livestock, will serve a host of important functions including marketing, education (breeders, members, and the general public), genetic and conservation management of the breed, and discussion of breed matters among the breeders.

Each member brings to an association certain strengths, needs, and perspectives. All of these must be managed effectively to assure that individual member needs are met, while also meeting the shared goals of the group. Each member likely has a slightly different priority for the different issues confronting the breed and the association, but all must work together for the common good of the breed as well as the association. Deciding which of these individual agendas fit within the association's job description is difficult and may sometimes be rancorous, but setting agenda priorities is essential if associations are to avoid diluting their efforts by attempting to be all things to all people.

Communication and education are important purposes for associations.

Purposes of Associations

Two important purposes for associations are communication and education. Most associations also operate a breed registry. Some breeds split the registry from the association and use the association more for management of shows, educational programs, and documentation of performance. Both approaches have merit and either can succeed. Management and promotion of the breed are at the core of an association's purpose, and these can be accomplished in a variety of ways.

Communicating the Association Purposes

Associations, networks, and even individual breeders all benefit from deciding on a purpose for the association. This is commonly called a mission statement. A mission statement should be a conscious effort to decide the reason for existence of the organization. Mission statements define goals and direct the actions of associations. Short, pithy mission statements are always best, as these are easy to keep before the members. They are also easy to use as yardsticks against which to measure decisions as they affect the breed.

A very telling passage from *Alice in Wonderland* comes to mind:
"Would you tell me, please, which way I ought to go from here?"
"That depends a good deal on where you want to go," said the Cat.
"I don't much care where," said Alice.

In the absence of goals, Alice fails to make progress. Courtesy of the Florida Center for Instructional Technology.

"Then it doesn't much matter which way you go," said the Cat.

The point is that a lack of direction nearly assures little to no progress. This is especially true of associations, and also all too true of individual breeding programs. Legitimate goals for associations, networks, and individual breeders are numerous. They certainly include breed conservation, management of populations for long term viability, commercial success in marketing products and breeding stock, increase in breed numbers, show ring wins, production of breed products, and production of breeding stock. The list is nearly limitless, but some of these goals can be mutually exclusive. Having a specified mission and goals helps to direct actions towards a successful outcome.

Membership

Breed associations usually have different levels of formal membership. Most associations insist that only active breeders (those registering livestock of the breed) can be full members with voting privileges. This is logical, because these are the individuals with the greatest stake in decisions and activities affecting the breed. Other associations open full membership to all owners, whether breeders or not. This last point is trivial to several breeds but is of great importance to horse breeds in which many owners are not breeders and never will be, but do have an important stake in many decisions concerning the breed.

Associate members usually are adults who do not own the breed, while junior members are youths. Both associate and junior members have some interest in the breed, but usually do not have voting privileges. Such members are indeed part of the association and can contribute to the discussion that goes into decision-making, but in most associations they lack a vote and so do not participate in actual decisions.

Family memberships are also a part of many breed associations, and usually

Youth involvement is important to the future of all breeds. Corey Drowns is holding a White Holland turkey. Photo by D. P. Sponenberg.

involve a financial savings over multiple individual memberships. Some associations limit voting rights to one vote per family membership, some allow two. A significant reason for family memberships is that those individuals covered by the membership can each conduct official business (registrations, transfers) with the breed association. This can greatly simplify registry and association function as delays are avoided when one family member is unavailable for signature on breed association documents.

Breed Associations for Endangered Breeds

Endangered and rare breeds bring with them a host of issues that are less important for common breeds. One important goal of associations that serve rare breeds is to try to locate and include all breeders of purebred animals into the association. It is also important to try to include all purebred animals into the registry, while excluding all nonpurebred animals. This is much easier to state than it is to accomplish, but this goal is critical for effective breed conservation. Associations benefit from having this as a stated goal kept before the members. Appropriate inclusion and exclusion of animals can be fraught with difficulty, and each association handles the challenge differently.

Breed associations for rare breeds must closely monitor genetic aspects of the breed.

Details of genetic management of breeds are especially relevant to rare breed associations. While associations for common breeds tend to focus more on genetic improvement of production, associations for rare breeds must also pay close attention to genetic aspects of breed population viability.

Communication

Communication is one of the vital functions of a breed association – and is probably even more important than registry functions for long-term survival of the breed. For rare and endangered breeds the communication network brought together by the association can be critically important to the survival of the breed and its expansion. Communication involves just about every aspect of breed business: news about breeders and animals, shows, successful promotion efforts, marketing, husbandry, and management. Communication is especially important concerning the dispersal of animals from herds that are being discontinued for whatever reason. Communication also involves education of all members as to genetic and political issues that face the association and the breed. The importance of communication can hardly be overstated.

Effective communication embraces a variety of mechanisms. Newsletters have long been a common method for breed associations to communicate with members and with interested members of the public. Newsletters are expected to be part of an association's service to its membership. The advantage of newsletters is that they stand as a reasonably permanent record of association activity. Small associations usually encounter problems in getting timely submissions of sufficient material for newsletters. This can be a major headache for an editor who is generally an unpaid volunteer. For many large associations the newsletter has evolved into a magazine format with a fully paid staff.

Communication Networks within Associations

In addition to formal associations, breeders also organize and communicate in many informal ways. These form what is loosely called "the network" and are a significant way to connect breeders to one another to share information and ideas as well as news of current interest.

Electronic mail or internet-based chat lists are becoming increasingly common as either official or unofficial forums for communication on topics of interest to breeders. These new electronic formats have the advantage of being quicker than print media, so that time-sensitive information is more likely to reach its desired audience than it is through more sporadic newsletters. Communication based on electronic mail is written rather than verbal, and it is fairly common for unintended inferences to be read into hastily composed messages. Another disadvantage is the more casual tone and use of electronic mail, which can result in significant misinformation or miscommunication unless electronic mail is wisely monitored and moderated. Unless misunderstandings can be rapidly defused they can escalate into nasty disputes. These are all too commonly encountered in electronic forums. If a significant subgroup within the chat group is dedicated to maintaining a positive and open discussion forum, an electronic chat group can be very successful for a group of breeders. For electronic forums to serve as conduits for official breed association business they must be closely monitored, and may indeed need be open only to members of the association in order for sensitive internal issues to be discussed and managed.

Associations work if breeders are engaged and involved in the breed's best interests.

One very powerful asset that networks bring to breeders is alerting them of emergency dispersals so that important portions of breeds are not lost. Networks have been essential in rescuing some specific strains of various breeds, and are an important part of the community that develops around a breed. This aspect is only likely to increase in importance and degree in the future as electronic communication becomes more widespread.

Unfortunately, gossip is also rapidly shared through informal networks and if not managed properly and promptly it can be disastrous to a breed and its breeders. Unsubstantiated allegations, or even accurate ones, can do irreparable damage not only to individual breeders but also to group integrity and loyalty. Public electronic forums are extremely poor avenues for addressing conflicts and rivalries.

Associations and networks can work only if breeders are truly engaged and involved in the best interests of their breed. All three – association, network, and

Randall cattle were successfully rescued from extinction because a network of people was able to assure they ended up with dedicated breed conservator Cynthia Creech. Photo by D. P. Sponenberg.

breeder – should be working for common goals, and the success of any one of these should contribute to the success of all three.

Multiple Breed Associations

Breeding of purebred livestock is generally undertaken by people that are passionate about the breed and its future. Breeders with an enhanced sense of this passion characterize the conservation of rare breeds of livestock. This passion can be a positive attribute or a hindrance to conservation depending on how it is managed. It is all too common for breeders of rare breeds of livestock to disagree on various points, and for those disagreements to result in splits of associations so that multiple associations end up serving single breeds. Splits within an association also split a small breed into yet smaller populations. Each of those populations is then more endangered than it once was as part of the previous larger whole. This is because small populations face biological risks from inbreeding, physical risks from disasters such as fire or storms, and also the very important fact that small groups simply have fewer resources for outreach, promotion, and education than do larger groups. Larger groups are much more likely to grow than are small groups.

Multiple breed associations serve a breed poorly.

Some of the disagreements that result in splits are significant because they originate back in the very concept of a specific pure breed and what it should entail. Most splits indeed take this form, and are usually the result of two conflicting philosophies. Generally one camp is more inclusive than another. No single answer fits this dilemma. Situations do arise in which the more inclusive breeders are failing to adequately conserve the breed. Equally, more exclusive breeders can easily be eliminating animals that rightly belong in the pure breed. These issues are dealt with in earlier chapters of this book, and an important key is for associations to develop philosophies that help them make these important decisions consistently rather than haphazardly. Moderating the extremes of opinion on this and other issues is a vital role for breed association leadership.

Much of the work of an association takes real effort. When multiple associations exist, all the efforts are duplicated. This includes registries, newsletters, legal corporate responsibilities, breeder recruitment, and a host of other issues. This aspect alone suggests that fewer associations are a better choice for most breeds. The duplication of effort drains energy and resources, and also detracts from other vitally important tasks such as breed promotion and broader communication among all of the breeders.

An example of divisions of associations are the breeders of American Quarter Horses in the first half of the 1900s. The breeders had several associations, each with slightly different goals and approaches. At some point leaders of several of these groups sat down and decided that it was all right to respect another breeder's horse and not mate your own horses to it. The result was to let the differences of approach coexist under a single umbrella. This, for a host of reasons, has resulted in the largest horse breed association in the world. One could argue about the genetic integrity of the breed, but it is undeniable that the association has taken this breed to a position that is enviable in modern horse breeds. And, importantly, foundation-minded breeders who favor the traditional foundation look and breeding are able to survive and function within the overall framework of a breed association that also includes a racing faction heavily influenced by Thoroughbreds. The foundation-minded breeders very much benefit from the resources of the larger breed.

A contrasting situation exists among Colonial Spanish Horse breeders. Splits among breeders of this breed have left it going by several names, including Spanish Mustang, Spanish Barb, Barb, Tacky, Florida Cracker, and others. This is a numerically small breed (regardless of which horses are included in the count) and currently has some 16 associations for fewer than 3,000 horses! Some of these associations are built around a single geographically defined population (Florida Cracker, Pryor Mountain, Sulphur). Others are more broad and inclusive (Spanish Mustang Registry, Southwest Spanish Mustang Association,

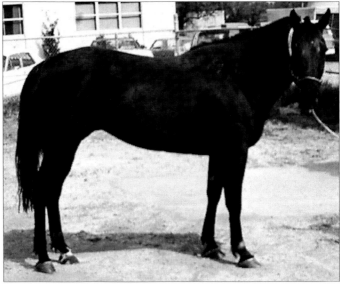

The American Quarter Horse has become the most populous horse breed through the actions of breeders willing to work together despite differences of opinion on certain issues. Photo by D. P. Sponenberg.

Horse of the Americas). Unfortunately, many of the splits occurred between strong personalities that were passionate about the breed, and those are the splits most difficult to heal. Each has good reasons for excluding what is excluded and including what is included. However, when the whole group of horses is examined they are generally much more like one another than like any other breed resource. In that situation, grouping these under a single association would help to foster conservation.

Probably unique in association history was a very inclusive effort in the establishment of the Jersey-Duroc (now simply "Duroc") Record Association in 1883. This association succeeded, and absorbed, other associations targeting red swine breeds (Victoria, Kentucky Red Berkshire, Red Guinea, Red Spanish, and others). The result was the composite Jersey-Duroc. The success of the Jersey-Duroc Record Association brought a kind of official extinction to many founder breeds of swine, but resulted in a remarkably successful breed based on all of them. The Duroc is noted to this day for its hardiness and productivity, and is currently one of the more common hog breeds in the USA. The success of the Duroc springs in large part from an effective and inclusive association.

Codes of Ethics

A Code of Ethics is a common way for an association to convey to its breeders their responsibilities to the breed as well as to the association. The usual minimum that most associations and registries insist upon is that pedigree and registration information be accurately represented by the breeder or other applicant for registration of animals. Various breeds, and various breeders, will add to this short version. Ethics can easily be called into question even when all parties are well meaning. The trick is to avoid penalizing the well-meaning honest mistake while penalizing intentionally dishonest actions. This is not always easy, and as codes of ethics become more and more complex it becomes increasingly difficult to bring the corporate wisdom of the association to bear in any given situation.

Codes of Ethics should only penalize dishonest action and not well-meaning, honest mistakes.

Some few associations are more active in the area of ethics and insist that business dealings by breeders and members meet certain standards of disclosure and honesty. Most of these guidelines and rules are an attempt to avoid customers dissatisfied by misrepresentation of animals. The identity and preparation of animals for competitive showing can be a frequent subject of codes of ethics, including the masking of birth defects or color markings in certain breeds.

Codes of ethics are sadly necessary, if for no other reason than that when many people are involved in associations (which is good), disagreements are sure to arise (which is bad). Expectations of breeders and members must be clearly specified. Penalties such as fines, limitations on registration or showing privileges, limitations on other association-related benefits, or outright expulsion must be articulated to correspond to the varying degrees of severity of breaches of ethics that can be encountered. Penalties must be wisely applied. They are certainly appropriate in some situations, but it is important to understand the effect of such punishments because they can quickly undermine any economic stability for affected breeders. Nasty disputes that end up in sanctions run a very real risk of alienating bystanders as well as the parties that are directly involved, and this can be very damaging to a breeder community.

Educational Programs

Ideally, breed associations are all involved in education. This aspect of associations is a key component to long-term success of the breed and the association.

One aspect of education for a breed association is the education of its own members and leaders. This is frequently overlooked. Lack of effort on this front can result in unfocused and unsuccessful breed associations. Exactly how to educate members is a tough question. All members, and especially new members, ought to have reasonably detailed knowledge of the breed, its unique traits, its history, and its historic and present function. This is asking a lot of anyone, but if all members have knowledge of these key points then discussions and activities can all gain a great deal of consensus and support from the mem-

Materials such as books and information packets can be very important for educating breeders. Photo by ALBC.

bership. Promotional pamphlets and breed standards can serve as educational tools internally, and establish a shared understanding of these topics among the membership.

It is especially important to educate members about the husbandry and management needs of the breed. This must take into account the historic role that the breed played, its current role in agriculture, and appropriate management techniques to assure breeder success. This is important not only for survival of the breed, but also for continued selection for historic type and performance. Much of this knowledge is being lost as older breeders pass away and new breeders with little or no connection to the breed, its heritage, and the culture in which it existed join the ranks. A resource list of publications, at the very least, should be available. If at all possible, interviews of older breeders and their children should document the culture and husbandry of the breed. Efforts like this can do a great deal to support future conservation of both type and function of breeds.

Breeder members must also be educated about the basics of animal breeding and breed maintenance. Many materials are available from ALBC and other sources (the ALBC *Conservation Breeders Handbook* is one such resource available to breed associations at volume discount). Assuring an educated group of breeders can avoid common pitfalls that detract from the survival of breeds.

Equally important are educational efforts aimed at those outside the breed association. These can include efforts to educate and enlist new members and new breeders to the association. This aspect is crucial to sustaining and expand-

Education of the breeders as well as the consuming public has been an essential component of the remarkable turnaround in the fate of Heritage turkeys, such as these Royal Palms. Photo by D. P. Sponenberg.

ing a support base for the breed. Failure here can easily lead to a diminishing base of breeders and enthusiasts for any breed. Promotional pamphlets, internet websites, strategically placed advertisements for the breed and association, and participation in appropriate public events are all useful to accomplish this.

Education of the general public about the breed and its products is also useful. Education helps to create a sound and consistent market. Assuring a secure place for breeds within production agriculture assures their future as genetic resources. An example is the ALBC's successful effort to educate the public about the desirability of traditionally raised standard and heritage varieties of turkeys. This effort has reversed the frightening downward trend of traditional turkey varieties so that they are much more secure now than a few years ago. The target was to link a genetic resource with public demand through education. This has been key to the survival and expansion of these varieties of turkeys.

Research

Forward-looking associations are interested in and involved in research. The level of research can vary from investigations into breed history to the analysis of molecular genetics of the breed. All levels are useful, even though at one extreme the studies are very inexpensive and at the other are tremendously expensive due to the technology, equipment, and personnel needed to succeed.

Historical research can help to direct conservation programs by putting a breed in its appropriate context. Landraces that are adapted to harsh local conditions, for example, can and should continue to be used and tested in similar environments. Breeds with certain unique uses (mule-producing jacks, for example) should continue in that niche. History can also help to identify the different bloodlines in a breed and help their conservation management so that none becomes extinct and the breed remains genetically healthy.

Many breeds also have a need for research on the inheritance of specific traits that are unique, such as color or functional characteristics. Research on the characteristics of the breed can greatly help efforts to conserve and expand breeds with attention to traditional types and roles of the breed. Research is also especially useful when deleterious traits surface in breeds. This research makes it possible to pinpoint the causes of the negative traits and to develop appropriate strategies for their management or elimination without endangering the genetic health of the breed.

Recruiting and Training New Breeders

Breed associations need to assure that new breeders are entering the breed. New breeders are the only mechanism for providing continuity of the breed, but their

American Mammoth Jacks continue to enjoy a specific role in the production of mules. Photo by Mary Ellen Nicholas.

recruitment is frequently overlooked as a deliberate and necessary activity for breed associations. Not only must new breeders be recruited, they must also be welcomed and then trained to be able to make critical selection decisions that conserve breed type, heritage, and utility. Failing in this is to fail in effective breed conservation.

In order for a breed to remain viable it is ideal that all breeders be seedstock producers. This means that all breeders should be producing animals that will be useful in purebred breeding programs. This broad base of contribution to the pure breed assures that no single breeding program, and therefore single bloodline, dominates the others and narrows the genetic base. For this ideal situation to occur, though, new breeders need to easily and openly be able to receive the fruits of the experience of more experienced breeders. Secret techniques for breeding superior animals all too easily pass into oblivion with the deaths of those breeders that zealously guarded them. Such secrets are notoriously slow to be rediscovered. Breeders need to set aside extremes of the competitive spirit, and instead need to take pride in watching the next generation grow in knowledge and competence. Effective mentoring of new breeders is not only satisfying, it is also invaluable to a breed's survival.

Developing the next generation of breeders has several important facets. One of these is conveying the cultural heritage of the breed, and this is where a rich history of stories about the breed and its breeders becomes useful. Stories such as boyhood deer chasing in the South Carolina swamps on the backs of Marsh Tacky horses enrich and inform the continuing use and selection of the horses.

The preferential use of four-horned Navajo-Churro rams in certain Navajo ceremonies similarly puts the maintenance of this variant in an important cultural context. All breeds have a heritage and culture that needs to be collected, safeguarded, valued and preserved in order to maintain the cultural relevance of the breed.

In addition to the more historical lore the fine points of selecting breeding stock need to be taught. The details of what to select, and why, are extremely important to convey to the next generation. For standard turkeys the selection of birds that are both large as well as sound and functional is a tough call, and the knowledge of how to do this is carried in the heads of poultry breeders such as Frank Reese who have long and effective experience in selecting birds for function in what has become a miniscule (but important) part of modern turkey production. The rarity of the system means that only a few breeders survive with the knowledge to make that system work. Only by having older, experienced, successful breeders eagerly and generously convey the subtleties of points of selection is it possible to assure that this rich knowledge is not lost. Transfer of this sort of knowledge usually takes personal interaction, for much of the detail is difficult to condense down to a written format. Breeder training short courses and field days can be very helpful in transmitting important details of breed selection to the next generation.

Selection techniques must be passed on to new breeders.

A great deal of knowledge and observation goes into successful breeding programs, and it is difficult to articulate and share some aspects of these. Breed type, for example, is more easily appreciated visually with live animals than by a written presentation. An appreciation for the interaction of breed type and breed function is also important, and needs to be conveyed to younger breeders so that the conservation of type has an appropriate and logical context rather than being a triviality of purebred breeding that appeals only to fanciers.

Established breeders have responsibilities in training new breeders, but new breeders also have a set of equally important responsibilities. New breeders need to cultivate a combination of attitudes and abilities. To successfully acquire the knowledge that is required to progress towards being a successful breeder it is ideal to have a passion for the breed, commitment to the long term success of the breed, adequate financial resources to manage and maintain a breeding population, a clearheaded commercial or utilitarian outlook that does not sacrifice breed type or heritage, personal integrity, an "eye" for good stock and for type, pride without arrogance, an ability to listen and learn from diverse resources, and to be reasonably free of assumptions.

Passion for the breed is critical for long-term success in the breed. Certain breeds "click" with certain people, and these are the combinations that work best. When new breeders select breeds, or strains within breeds, it is important that they find a project that inherently appeals to them. Passion is difficult to force into a situation where it is not already present, and passion about a breeding program is a key component for the dedication that is necessary for success.

Commitment is important because breeds benefit from long-term programs much more than they do from short-term endeavors. It is almost invariably damaging to breeds when breeders acquire stock and then disperse it all after only a few generations of breeding. While dispersals occur for a variety of reasons, a committed breeder will work diligently to place key breeding stock in the hands of committed breeders rather than dispersing them indiscriminately or sending them to slaughter. Similarly, informing family and friends of wishes for a herd's dispersal may be a final gift to the breed that a breeder has invested so much in. Breeding programs can only really contribute to breed maintenance if they take a long view rather than a short one, and commitment is key to this.

Passion is essential for success.

Finances must be adequate in order to maintain breeding stock. In most cases the products of the breeding program should cover the cost of maintaining breeding stock. Some people, however, find it difficult to deal with outright commercial aspects of breeding most breeds of livestock. When it becomes impossible to send excess animals to slaughter or to other breeders, then animals accumulate and numbers become greatly in excess of what is needed for breeding programs. These excess animals do not contribute to effective breeding programs, nor do they contribute to positive cash flow that is needed to maintain breeding stock. This can have a decidedly negative effect on a breeding program. This is a very important aspect, and must be carefully considered before embarking on a breeding program for any breed.

Most breeds benefit from a practical commercial approach that also acknowledges breed type and breed heritage. An extreme commercial approach can ignore breed type, and this results in changing breeds away from breed type rather than a more subtle and productive selection within the constraints of breed type for commercial utility. Morgan horses, for example, have a number of bloodlines that are a park-horse type rather than the earlier, traditional multipurpose farm horse type. This change was largely driven by market forces and breeders willing to sacrifice breed type for commercial gain. The uniqueness of the breed was eroded by this, and the result has not been an effective strategy for breed maintenance.

Breeds benefit from a commercial approach.

Moral integrity of all breeders is necessary to safeguard the reputation of the individuals, the association and ultimately the animals themselves. Records must be honest, and animals must be honestly represented to both registries and customers.

Developing an "eye" for breed type and good stock is a subtle but important ability that master breeders possess. For many people this is almost instinctive, and therefore difficult to describe. It is even more difficult to train other people in how to achieve "eye," but eager young breeders can go a long way by seeking out older breeders and not only talking to them, but also trying to see with their eyes when inspecting animals. It is especially helpful to inspect and evaluate animals along side an old established breeder, because this activity brings many subtle but powerful details to bear. Livestock judges can help greatly in this by inviting young breeders to evaluate livestock alongside them, asking the young breeders "what do you see?" rather than telling them up front what the experienced breeder sees. Forcing younger breeders to actively inspect and evaluate stock is an essential component of evaluating an eye for type and evaluation.

Developing an "eye" for stock is important.

Taking pride in the fruits of a breeding program without being arrogant is also important. Most successful breeders are reasonably unassuming, and listen effectively to others. They can sort through what is said, and can learn something from just about anyone. They put this stored knowledge to good use. Arrogance precludes the receipt of much information, and defeats many breeders because they are missing important bits of information or technique.

Becoming a master breeder is slow, complicated work, but deliberate cultivation of these abilities and attitudes by older breeders and associations can help greatly in the process. The greatest master breeders have all of these attributes, but these true masters are few indeed. Nevertheless, knowing what is needed for the breed to carry on is important to associations. Knowing what is needed can enable associations (and breeders) to identify and nurture young breeders with potential to succeed. Both associations and experienced breeders can help young enthusiasts understand the level of personal commitment needed for the personal and intellectual education they need in order to master the breeding of livestock or poultry.

Becoming a master breeder is slow work.

Breed Promotion

In the end, promotion is a key component of every successful breed association education program. Breed promotion is multifaceted, and involves educating breeders as well as the general public about the breed's products and potential. The goal is to encourage and promote demand for the breed. This can be demand for a traditional free-range turkey for a special Thanksgiving Day meal, or a tasty spring lamb roast from a Tunis sheep. Brown eggs from a Dominique hen, or the wool from Leicester Longwool sheep that offer handspinners and other craftspeople exceptional raw material for their endeavors are other good examples. In addition there are valuable services or functions of the breed such as the crossbreeding potential of Florida Cracker or Pineywoods cows or the remarkable brush clearing abilities of Tennessee Myotonic goats. Each time the purebred product finds a market, purebred breeders benefit and so do the breeds they steward.

Not only must the breed excel in some special niche, the word must also get out so that those interested in the product or service know where to get it. The goal of promotion is to accurately portray the breed's excellence so that a loyal and satisfied following is established. Breed promotion is only successful if it is based on facts that reflect the breed's uniqueness and superiority.

The goal of promotion is to portay the breed's excellence.

Promotion must also reflect the breed and its identity. For this to be successful the association must have clear knowledge of the breed and its history. By being firmly rooted in this knowledge it is possible for the breeders and the association to value the character and function of the breed, and to promote these attributes in the marketplace.

Breed Sale Events

Many breeds have annual sale events. These sales serve as a promotional tool that brings breeders, buyers, and other interested folks together in a setting where animals are available for viewing and purchase. Most of these breed sales are auctions, and are an opportunity to get started with the breed.

Breed sale events work best when old, established breeders support them by consigning select animals so that the sale reflects the top end of the breed. This is one situation where it is vitally important for breeders to realize that the good of the entire breed takes precedence over the short-term benefit to individual breeders. Especially in the early years of such sales it may well be that the prices are somewhat below those possible by private treaty sales. A multi-year commitment to assuring high quality lots for sale can help to assure that the long-term fate of these sales is good and strong.

Sales committees must be visionary, hard working, and neutral in order to assure that the sale is run fairly and is successful. For many breeds it may be necessary to constrain consignments of breeders. For example, large numbers of breeding males at a sale can easily lead to an overall weak sale, with numbers of animals going for slaughter instead of for breeding. This reduces the prices to commercial levels, and is not in the best long-term interest of a breed sale event. A strategy to circumvent this problem is to only allow breeders to consign males if they also consign three or more females to the sale. Specific invitations (or requests) to larger breeders for consignments of select females will generally assure a successful sale.

Forms of Association

Associations must be organized, and while organizational details usually take only a few different forms, each association ends up being individual in exactly how it is developed and organized. The forms of organization have legal ramifications. Among small groups the legal aspect of association is usually not considered terribly important, but as associations become larger and interactions become less personal, it is vital for associations to function as legal entities. This prevents an association's legal issues from spilling over and becoming personal legal issues for the members and leaders involved.

Private Associations

One mode of organization that is rarely used among breed associations is private ownership. These can work very well, although by the very nature of their organizational form they are less participatory than other forms of organization because the owner is the final authority for all decisions as well as for all financial responsibilities. This can have advantages as long as the owners are dedicated to the advancement and persistence of the breed. This is usually the case, so that the few private breed associations do indeed function well for the advancement of the breeds involved. One significant weakness of private ownership, though, is continuity of services and philosophy when ownership eventually changes. The transfer of the association from one owner to the next must be carefully considered, and because this is a private rather than a public transaction few rules exist to guide this process.

A few examples of privately held associations illustrate some of the issues with this form of organization. The American Indian Horse Registry has functioned for several decades as one of the registries for the Colonial Spanish Horse. Sole ownership passed decades ago to a new owner through outright purchase, providing for continuity albeit with a subtle shift in emphasis towards more

The American Indian Horse Registry has emphasized the conservation of traditional breeding in the horses, such as this Choctaw strain Colonial Spanish Horse. Photo by Bryant Rickman.

traditional bloodlines within the breed. This is a boon to conservation, although the opposite could have easily happened as owners get to decide what to do with private associations.

Another registry for Colonial Spanish Horses is the Horse of the Americas. It was passed to new owners through purchase, and a more participatory form of governance is in place through the use of directors, each representing one of the strains of this important landrace. These people generally have a relatively high personal stake in the breed and its future, and so are engaged and active in breed activities.

The Myotonic Goat Registry was recently founded to provide registry services for Tennessee Myotonic Goats. Though a sole proprietor organization, the owner has designated a board of advisors who consult on issues of breed maintenance and registry procedures. No decision is taken without the consent of the advisors. This helps to provide broader insight and thinking, especially for potential problems, than is possible when any one person (however capable) makes all decisions in a vacuum. The advisors periodically invite other people to join them, resulting in a broader voice in breed issues and also assuring that the various segments of the breed are represented. Continuity is assured by appointment from within the board rather than by election from registry breeder members, which may strike many as very undemocratic. It is indeed undemocratic, but it is a strategy for providing stability, some innovation, while also assuring a decreased opportunity for some of the nastier political upheavals that can beset smaller breed boards that have significant turnovers following elections. These upheavals, however well intended, almost invariably result in great

Tennessee Myotonic goats have benefited from association continuity. Photo by D. P. Sponenberg.

and lasting damage to the association. Private associations circumvent most of that, but do run a risk of not enough responsiveness to breeder constituents. Some of the above strategies are good compromises to accomplish continuity and responsiveness.

Unincorporated Associations

Sometimes a group of breeders begins networking and acting like a breed association. These casual associations can be very effective in breed conservation. In the interests of the breed it is usually useful for these networks to adopt a formal structure in order to assure continued success. Following the guidelines for an incorporated association is a good plan from the earliest days of group activity because this makes future changes less challenging.

Incorporated Associations

Members organize most breed associations as corporations managed democratically with participation. This organizational style can clarify the tax status of the association, as well as issues of personal versus corporate liability for the board and officers. Corporations are usually chartered at the state level, where laws vary as to the legal requirements and costs that must be met to constitute a legal corporation. As a result of the state-to-state variability in laws governing corporations, it is nearly always necessary to have local legal assistance in incorporating a breed association.

Breed associations that choose to incorporate can do so as either non-profit or for-profit corporations. The distinctions have tax and legal ramifications. While the non-profit status does have some advantages, this status has more complicated reporting requirements that may make this choice a poor one for many smaller associations. One advantage of non-profit status is the ability to obtain grants from certain sources as well as tax-deductible contributions that are unavailable to for-profit organizations. The for-profit associations circumvent some of the reporting procedures, but have other disadvantages such as ineligibility for certain sources of funding. It is rare for associations to be making enough money to truly be considered profitable, especially for rare breeds, so that non-profit status usually makes good sense for most of them.

Bylaws

Bylaws dictate the organizational structure of the association, and must be tailored to meet legal requirements for associations, especially if the association is incorporated. These requirements vary from state to state, and so checking on the requirements is a case-by-case situation. Bylaws of associations must be carefully developed, because these dictate the management of the association in terms of leadership, selection of leadership, terms of office, financial responsibilities, and management of any association business such as shows, registrations, meetings, or other activities. Bylaws are generally accepted by vote of the membership, and contain procedures for their amendment. These are important details, for they govern the legal aspects of the association. Bylaws need to be stringent enough to provide for continuity and consistency within the association, but also need to be flexible enough to allow changes when necessary. Procedures for changes need to be practical and not overly cumbersome. Members should have easy access to the bylaws so that the association is sure to always operate within their guidelines. Guidelines on developing and revising bylaws are readily available in various books. Discussion with other breed associations might also provide useful guidance.

Bylaws organize the association.

Procedures for amending bylaws are extremely important, and need to be explained and codified in the bylaws. As associations change over the years, and especially as membership either grows or shrinks dramatically, it usually becomes necessary to amend bylaws to assure that association functions can continue in an efficient manner. For changes to be legitimate the process by which they are made needs to be clearly explained and followed. In most associations changes can be put forward by either the membership or the board, and then usually must be approved by vote of the entire membership. This can be either at an annual meeting (at which only a portion of the membership is likely to attend) or can be by ballot from the entire membership (which is at least potentially more inclusive and therefore offers fewer opportunities for accusations of a deal done behind closed doors). In the case of a non-profit association, changes in bylaws need to be recorded with the state granting the non-profit status.

Board of Directors

Incorporated associations are required to have a board of directors. Commonly the board of directors is elected from the membership. Voting for the board of directors is frequently restricted to breeder members, and usually excludes associate members who are either youth, non-breeders (for most species) or non-owners. Some horse breeds allow non-breeder owners to vote because they constitute a significant portion of membership and have a direct interest in decisions that affect showing and competition. A few common variations include a division of the board into regions or districts so that relatively uniform representation is assured to all members across the nation. Usually each district or region has a designated number of positions on the board, and is able to fill those by direct election.

The directors are usually elected from the membership at intervals established in the bylaws. A common strategy is to have an odd number of directors to avoid tie votes, and for the directors to serve terms of multiple years. Three-year terms are routine across many associations. In most associations these terms are staggered so that roughly one third of the board members are elected in any given year. For new boards this would mean one third of the directors would serve a one-year term, one third a two-year term, and one third a three-year term. In each succeeding year each director would serve a three-year term. The result of staggered three-year terms is greater continuity of leadership for the association than would be the case if all director terms ended and started at the same time. Term limits are typical in most associations, and the most common is a limit of two consecutive terms.

Board members usually have staggered terms.

Changing board membership brings new dynamics that serve to avoid stagnation. A potential downside to these periodic changes, especially with small boards, is that a large and sudden shift in leadership can be disruptive to the continuity of services and association culture. In some situations board "takeovers" have been instigated and successfully accomplished by replacing several board members through the normal election cycle. While this is indeed democratic, and can bring desired and much-needed change, sometimes the change is so rapid and acrimonious that the association suffers loss of members and breeders, as well as some loss of loyalty from those who remain. Any sudden change in direction or philosophy needs to be carefully considered and accomplished in order to avoid doing more damage than good.

Boards of directors vary in their activities and responsibilities. Some of this variation is due to differences in bylaws, while other variation is simply due to differences in association culture. For most associations the directors provide leadership and are especially crucial in developing and maintaining policy. In larger associations it is generally the board of directors that establishes policy, and oversees the executive director. The executive director is responsible for establishing, maintaining, and changing procedures, which implement the policy. The executive director also oversees the rest of the paid staff.

Meetings of the board of directors always have legal requirements for a record of the proceedings in the form of minutes. Board procedures must also be spelled out so that orderly meetings are possible, usually following *Robert's Rules of Order*. Fortunately these rules are the norm for many livestock organizations, and most people have at least some familiarity with them. They are readily available in book form.

Directors and Officers

Members of the board of directors are ultimately responsible for the fate of the association, and in most states that also includes the fiscal fate of the association. Beyond their legal responsibilities, members of the board usually have important informal responsibilities by setting the tone for the entire breed association. If board business is accomplished in a businesslike and open manner, the association benefits. If board members are self-serving and manipulate the association through secret deals, then the association is diminished. Setting the tone of the overall association and its activities is one crucial responsibility of the members of the board.

Setting the tone of the association is the responsibility of the board.

In addition to the formal responsibilities and activities that the board of directors has by virtue of the bylaws, they are also responsible for a great deal of what works well about associations. They must set an example to the general membership by eager participation in the association and its activities. Some boards function strictly as a governance body, while others are involved in the day-to-day functions of the organization. The former is more typical of a mature organization while the latter typifies many small or young organizations. Ideally members of the board volunteer for a variety of responsibilities such as planning and preparing for meetings, sharing their knowledge and experience with members through newsletter and other communications, and also contributing key abilities to the association such as specialized training (legal, veterinary, fiscal, writing, design) that they might have to offer to the association.

Networks of Breed Associations

A few organizations are available that provide resources that help to consolidate and inform the individual associations. One of these is the American Livestock Breeds Conservancy, which is especially helpful in organizing and providing information for rare breeds and the associations serving them. Other consortia of herdbook associations exist, such as the National Pedigreed Livestock Association, however most of these target larger breed associations. These larger consortia frequently engage in political lobbying and other large-scale functions.

Some breed groups are organized along species rather than breeds. These include groups such as the American Rabbit Breeders Association (ARBA), the American Poultry Association (APA), and the Society for the Preservation of Poultry Antiquities (SPPA). Many of the more populous poultry and rabbit breeds also have stand-alone breed clubs, but still depend greatly on the ARBA, APA and SPPA for organizing and disseminating information and activities of interest to the breeders.

Species Associations

Very few associations exist that are active across an entire species. Cattle breeders are served by the Dairy and Beef Councils, as well as the state and regional cattlemen's associations, although these are much more involved with the economic dynamics of these industries than they are the issues of purebred breeding and breed maintenance. The same is true of the Pork Council, the American Sheep Industry, and the National Turkey Federation.

Rabbits enjoy the benefits of a single umbrella association for all breeds. The Giant Chinchilla is an excellent meat breed. Photo by Jeannette Beranger.

In contrast, dairy goats are very much monitored by associations (American Dairy Goat Association, American Goat Society) that supervise multiple breeds. In this situation the breeds benefit from sharing expenses for registry functions. This is a very real boon to the numerically smaller breeds, as they are still able to maintain timely support for registry functions. A possible negative, however, is that underlying philosophies become shared across breeds rather than being tailored to each individual breed. The result is that some blurring of breed distinctions has occurred over years of selection under a single guiding philosophy.

Promoting the Association

Promotion is very important if the association is to secure its rightful place as the hub of breed activity. Active promotion is best if built upon a base of association integrity and competence, but these two in the absence of active promotion will be insufficient to ensure the success of an association.

Association Reputation

The reputation of the association is based on many different aspects that reflect the complexity of an association's work and responsibility. One aspect of an association that is critical to its reputation is commitment to the breed and bylaws. Only a sincere and clear-headed dedication to the breed and its maintenance can enable an association to convey the message of the breed to the public. Bylaws must serve this mission, and the association must adhere to them in an honest effort to secure the status and future of the breed.

The reputation of the association is also earned by its service to members. This is not always an easy job description. Associations must serve as hubs of information and services, and must be available to offer these to members. This is more easily accomplished by large associations that have central paid staff. It is much more difficult for smaller associations that are staffed by volunteers, for these volunteers will be the recipients of many requests for information and services on behalf of the members as well as the public. While some members can be unrealistic in their demands on even the largest and most well-staffed association, it remains true that reasonable requests must be acknowledged and met efficiently and pleasantly if the association is to attract and retain members. The cost in time, effort, and financial resources is significant, especially for small associations, and all of these need to be understood and considered by volunteers and by the board as it develops budgets.

The reputation of an association is earned.

Association Responsibilities

Associations have various responsibilities to breeders as well as to the breed. These are usually congruent, but occasionally it is possible for the breeders' interests to be at variance with those of the breed. It takes dedication and neutrality for the association to assure that the needs of the breed are met while not damaging the situation for the breeders. Difficulties arise when short-term commercial interests conflict with the long-term survival and integrity of the breed. Examples are changing fads in type or purpose of the breed, where a new or off-type may be in temporary high demand. These can result in significant economic benefit to some breeders, but can also irrevocably change the breed type.

Conservation Responsibilities

One obvious responsibility of the association is to provide for education about purebred breeding and breed characteristics to current and potential breeders. This needs to be a very active role of the association, for to neglect this is to assure that less organized or less informed sources will flow into the vacuum left by association inactivity. Without a well-thought and articulated stance on conservation and purebred breeding the association and the breed are both likely to drift from fad to fad.

Providing Pedigree Information

Associations generally maintain complete pedigree records for their breed. This is important information to breeders as well as to others interested in the breed. Pedigree information is valuable, and it is worthwhile for each association to actively decide to whom pedigree information should be made available and under what conditions. Some breeders may feel that pedigree information on animals in their ownership should be privileged. Others will have no disagreement with free, open, and total sharing of such information with any and all who request it. Breed associations need to decide which of these approaches will be the procedure to follow.

It is generally in the breed's best interest for pedigree information to be widely available.

It is generally in the breed's best interest for pedigree information to be widely shared and available. In most associations a nominal fee is often charged for pedigree information.

Providing Breed Health Status Reports

Associations should be aware of the general health issues confronting the breed and should communicate these to breeders and members. An important aspect of genetic health is the occurrence and incidence of genetic defects within the breed. The reporting of specific genetic diseases can become politically heated, and it is best if the information is stripped of any identity linking it to specific breeders or animals. Anonymity comes closer to assuring full disclosure than does any other strategy.

Although anonymity does have the drawback of limiting the actions that interested people can take to select against specific defects, it does assure that breeders are not carelessly nor needlessly smeared. It is highly likely that misinformation will sometimes be conveyed. If an anonymous reporting scheme is actively pursued it generally develops that owners and breeders step up to the challenge of disclosing information more publicly. This is especially true of defects for which testing is available and which are increasing in number all the time. In a situation where animals can be DNA-tested as either free of the variant or as carriers, it is more and more common for that information to be made public by breeders so that customers can make informed choices.

Any health or genetic information disclosure should come with education about how to use the information. For example, in many rare breeds the draconian measure of culling all males that carry certain recessive traits or structural defects is likely to assure the constriction of the gene pool and the eventual demise of the breed. In this case the breeders need to be educated as to the wise (and necessary) use of carriers, and how to do this with minimal risk of expressing the undesired traits or defects.

Useful examples occur in a number of breeds. One is the chestnut allele in

Cleveland Bay breed association rules penalize chestnut foals as unregisterable, but programs to avoid these must be carefully crafted in order to not diminish a rare breed even further. Photo by Jane E. Scott.

Cleveland Bay or Friesian horses. Chestnut foals are "off standard" for these breeds, and therefore breeders try to avoid producing them. In the past, a small percentage of such foals were born. With the advent of DNA testing for the chestnut allele it is possible for breeders to know which animals are carriers of this allele. With that knowledge comes the temptation to cull from breeding all carriers, or at the very least all stallion carriers. What is important to remember, though, is that the carriers are very likely to represent a single or a few bloodlines. By removing all carriers (or even all carrier stallions) it is easily possible to remove an entire bloodline from a breed that needs all bloodlines to assure genetic diversity in the breed. A safer approach is simply to avoid mating two carriers to each other, and thereby eliminate the risk of any production of chestnut foals. DNA test results can easily be used to accomplish this end.

While color is a somewhat trivial and cosmetic example, some examples include heritable diseases. Availability of DNA tests for these is increasingly possible, and raises the issue of what to do with such information. In some rare breeds the elimination of all carriers runs the risk of eliminating too much genetic variation. A safer approach is to insist on testing, but to then use test results to not eliminate carriers as breeding animals but rather to constrain their mating to noncarriers. This strategy avoids the production of affected offspring, but also retains the broader genetic value of the carrier animals. It is vitally important to remember that culling removes entire animals from breeding. Culling removes not only the single targeted undesirable gene, but genes that govern many desirable traits as well.

A troubling example of genetic testing and breed maintenance is the HYPP (hyperkalemic periodic paralysis, or the "Impressive Syndrome") trait of horses. This is due to a dominant gene, and results in heavily muscled horses that can occasionally collapse and can die. While Quarter Horse breeders have been diligent to quickly identify and reduce the incidence of most lethal traits, this one has persisted in the breed. A major reason for its persistence is that many HYPP horses are consistent halter class winners. To correct this problem it is essential that the show ring and the breed association rules speak with one voice on the issue of genetic flaws. They should both diligently assure that animals with flaws do not meet with increased demand, for in that situation it will be impossible to reduce or eliminate the flaws even in the presence of genetic testing. Market demand ultimately will dictate the fate of genetic defects, and in nearly all cases the situation needs to be that the defect carries with it a market penalty.

It is essential that the show ring and the breed association speak with one voice on genetic flaws.

Rare variants, such as polled Pineywoods cattle, are important to monitor in order to assure they don't slip into extinction. Photo by D. P. Sponenberg.

Reporting Measures of Genetic Diversity

The genetic status of a breed is one dimension of its overall genetic health. Association actions in this regard should include periodic reports on the status of bloodlines within the breed, overall population levels, popularity trends within the breed, and other trends that are important to maintaining the breed as a viable population. Incidence of specific genetic variants (rare variants such as polledness or dwarves in some breeds, for example, or color or gaitedness for some horse breeds) should also be assessed and communicated to the breeders on a periodic basis. Periodically reporting these measures, such as two-horned, multihorned, or polled rams in the Navajo-Churro sheep breed, assures that breeders are alert to variants that could otherwise slip to extinction.

Development of Programs to Save Herds in Peril

Action plans for rescuing herds that are in peril of being disbanded through commercial (and this generally means slaughter) channels are a necessary responsibility of the association. Programs to fend off sudden loss of genetic material need to be articulated in advance of their need, for the need generally occurs after a catastrophe (sudden disability or death of a breeder, natural disaster) such that quick and informed action is needed without the luxury of time to develop effective strategies on the spur of the moment.

Programs to avoid loss of genetic material must be in place before an emergency requires them.

Associations can help to build a breed culture of informed awareness concerning dispersals. This should ideally result in each breeder having a plan for the dispersal of his or her herd in case of accident or other disaster. Association plans should very much be only a last resort to use when breeder plans fail.

Development of Long Range Conservation Plans

As a critical member of the stewardship team for a breed, associations should actively develop mechanisms for the long-term maintenance of the breed. These involve all factors affecting breed survival. Included are economic parameters such as products, their marketing, and the association helping to assure a brisk demand for the purebred product. "Certified Angus Beef" is a very good example of a breed association undertaking effective breed product marketing and reaping a huge reward from increased demand for the breed. The marketing of heritage turkeys for Thanksgiving is a similar success story, and one that needs to be repeated for every breed.

Also included are long-range breed maintenance plans that detail the association's actions to facilitate the maintenance of the breed as a genetically viable and useful entity. Conservation plans should be based on accurate assessment of the status of bloodlines within the breed. Plans can be formulated for assuring that different bloodlines survive throughout the breed. Plans to cryopreserve appropriate genetic materials for long-term storage should be implemented and should involve targeted sampling to assure the broadest possible representation of the breed. Special attention should be paid to the inclusion of an adequate sample of all distinct bloodlines within the breed. This can be accomplished by individuals but will be more effective if coordinated by the breed association.

Heritage turkeys enjoy a secure future due to demand for purebred products. This Slate tom is both beautiful and tasty. Photo by D. P. Sponenberg.

Participation with the National Animal Germplasm Program can provide technical support and other resources for this endeavor. The ALBC is a partner in this program, and can help coordinate breeder activities with it.

Dispelling False Rumors Quickly

Associations are, by their very nature, groups of people. Groups of people always involve politics as well as other social dynamics. Rumors tend to be part of the whole package of a breed association, but ideally false rumors can be dispelled quickly and effectively by the board and other leaders within the breed. More effective yet is for associations to build an open, accurate, and effective communication style that prevents any need for a rumor mill. It is especially the responsibility of breed leaders, such as the board of directors, major breeders, and the staff, to very deliberately avoid contributing to false rumors and negative behavior. This sets a positive tone for communication among the breeders, and is of great benefit to the entire breed and its community of breeders and consumers.

Conflict of Interest

Certain conflicts of interest are nearly inevitable in a breed association, and it is wise to deal with these before problems develop. One source of conflicted interest occurs when the registrar or secretary is also a breeder. This is a very common state of affairs in small breed associations, where funds are not available to outsource the registry function.

Most breed inquiries are directed through the secretary or registrar, which is why a breeder in such a role can have an inherent conflict of interest. It would be easy to skim off purchase inquiries for oneself, or for close associates, friends, or partners. To prevent even the appearance of this, it is wise for a specific packet of information, including members' and breeders' names and addresses, to be forwarded to all inquiries. This strategy allows the inquirer to then take the next step of contacting breeders in a specific area or with specific animals of interest. In the interest of fairness to all it is always wisest to forward complete information to any inquiry. Some larger associations limit information to that of the same region as the enquiry, but it is always safest to forward reasonably complete information so that any impression of favoritism is prevented.

Another source of conflict can occur with boards of directors or other officers that are responsible for making breed and registry rules. Such rules can degenerate into protecting self-interest unless association members are careful to establish an association culture in which breed interest supersedes all other interests. Specific examples of shortsighted rule making include changing the exclusion of cryptorchidism as a disqualifying defect in the stallions of some breeds. Board members had stallions that were cryptorchid, therefore they contributed to self-interest that over-ruled long-term breed welfare. A similar conflict of interest in some landrace populations has resulted in early and complete closure of herdbooks, thereby eliminating many purebred animals from the registered breed. On the other hand, at least one landrace herdbook included a few known crossbred animals because prominent individuals had already heavily invested in those animals before their crossbred character had been realized. These examples are instances of short-term self-interest overriding the breed's long-term welfare. All breed associations should be diligent to create an association environment that engenders a high regard for the long-term integrity of the breed and itself.

Local and Regional Groups

Some breed associations have regional or local subdivisions, especially if the association is large. The regional and local groups may have formal status, but most often are groups that arise to allow for greater local communication and activities such as shows and meetings, field days, and other get-togethers. Developing a formal status for these groups can assure that all breeders are speaking and acting in accordance with the philosophy and standards of the association.

7. Competitive Shows

Competitive showing has long been associated with breed associations, and indeed most of the breed association movement a century ago revolved around public competitive showing of livestock. Showing of livestock has both negative and positive consequences for breeds, and these tend to be accentuated in the case of rare breeds.

Positive aspects of competitive showing of livestock include public exposure. Shows are a potential source of new breeder recruitment. This is especially true if the showing is seen to be fair, fun, and productive. Another positive aspect of showing is that the placement of animals in rank of quality can guide breeders as they make breeding decisions. Showing and judging can have a profound impact for keeping breed type intact, but only if show ring evaluation emphasizes breed type. This can be especially beneficial for educating new or young breeders.

Competitive showing can also have negative consequences for breeds and breeders. If judges evaluate multiple breeds by the same mental picture of excellence, showing can take a breed away from its traditional type. A single mental image of excellence ignores or blurs distinctive breed type that is so critical in maintaining breeds as distinct genetic populations. Judges can easily damage breeds by ignoring off-type characteristics in animals placed well up in the winnings. Changing of type can be a result of show ring and judging fads, and is detrimental to a breed and its genetic integrity.

Showing of livestock has both negative and positive consequences.

Competitive showing tends to select extremes as ideal, rather than favoring more balanced animals. Favorable placement of extreme animals has the effect of driving selection towards those chosen extremes and away from moderate and balanced animals that may well prove to be more functionally useful. An unfortunate aspect of the show ring is that anyone, including trained judges, tends to be drawn to an extreme animal as a "first impression" response. This is easiest to appreciate in dog shows: in toy breeds the smallest makes the first exceptional impression, while in large breeds the largest dog often does. The

Competitive showing was one of the original purposes for many breed associations and remains a popular activity today. This is a Navajo-Churro show at Dine College in Arizona. Photo by D. E. Bixby.

result has been, for many breeds, that small breeds continually become smaller and large breeds continually become larger. This is frequently to the detriment of overall balance, function, and soundness.

Show ring wins can translate into significant monetary earnings. This opens the door for a "win at any cost" mentality that can be very damaging to the breed as well as to breeder integrity. The show ring can end up dictating most breed policies for these breeds. When this occurs, other weighty aspects of breed biology and maintenance can be ignored to the long-term detriment of the breed. While any breed is made up of individual animals, it is a subtle concept that those individuals all must be evaluated relative to their potential contribution back to the breed as a gene pool. An example of the show ring subverting breed integrity occurred a few decades ago in the Argentine Criollo horse.

Competitive showing tends to select extremes.

The traditional horses were small and tough, but show wins began to go to taller horses. The breed increased its height but lost some of its ruggedness in the process. Fortunately breeders recognized what was happening in time to correct the judging error, re-emphasized the traditional Criollo type, and were able to pull the selection pressure back in a more traditional direction. A similar phenomenon has occurred within the Texas Longhorn cattle breed in which horn length, color, and size became emphasized over longevity, production, and hardy adaptation. Some breeders, as a reaction to this trend, have gone back to deliberately favoring the traditional type.

Show ring strains of Texas Longhorns have undergone a change in type as a result of show ring practices. Photo by D. P. Sponenberg.

Card Grading

Card grading, or evaluating individual animals to the standard, began in the United Kingdom as an attempt to make sense out of huge classes of poultry that were presented for evaluation and competition at some of the large shows. Card grading has been further developed by the Rare Breeds Survival Trust in the United Kingdom, as well as the American Livestock Breeds Conservancy in the USA. It has evolved into a system that offers many benefits to offset the potential negatives in traditional competitive showing.

Card grading involves evaluating each animal according to the ideal of the breed standard. This contrasts with competitive showing in which the animals are evaluated against one another. Card grading is usually accomplished by three judges to avoid a tie. The animals are each evaluated separately by the judges and in the absence of the owner or handler. The judges examine the animal (which includes hands-on evaluation), and should also have it walk freely up and down an alleyway. Watching the animals move is very important for all breeds, but can be tricky to manage for poultry evaluations. The animals are placed into categories: excellent breeding stock (blue card), good breeding stock (red card), acceptable breeding stock (yellow card) and unacceptable for breeding (white card). In the United Kingdom, the colors are reversed for the first two

Card grading involves evaluating each animal according to the ideal of the breed standard.

categories so that a red card indicates excellent breeding stock while a blue card is given for good breeding stock.

Card grading has significant advantages over competitive showing. Each animal that qualifies for evaluation on the basis of the appropriate documentation (registration and identification requirements) is given a card reflecting its overall quality with respect to the breed standard. No specific proportions of the classifications are set, so that it is possible to have all animals receive blue cards at a given event, or, all receive white cards. The usual result is a handful of blue cards, relatively more red, and then varying numbers of yellow and white depending on how selective breeders were in bringing animals to the event.

Card Grading Categories are:
- *excellent (blue)*
- *good (red)*
- *acceptable (yellow)*
- *unacceptable (white)*

The general trend over years is that the higher categories (blue and red) become better represented than the lower ones, as breeders become educated in evaluating their animals with respect to the breed standard. Breeders then leave the potential white card animals at home rather than bringing them for public evaluation. In this regard, card grading does a great job in educating breeders about breeding to the standard.

Card grading sends a very powerful signal to breeders that no single animal is best. This is important, for it allows novices to realize that several different animals, each with a specific array of strengths and weaknesses, can be of equal value to the breed. It also works to educate breeders that not all animals have equal value to a specific breeding program as each has unique strengths and weaknesses. Once the elements of card grading are understood, breeders can apply this same evaluation to their animals at home when selecting breeding stock.

When card grading is accompanied by an open and public "reasons session" it is an especially powerful mechanism for breeder education. Judges' comments concerning the difference between an "overall weak" yellow card animal and a "single major flaw" yellow card animal, for example, can alert breeders that each animal must be evaluated individually, for each has a different and appropriate role in the breeding structure of a herd or breed. In this example, an "overall weak" yellow card animal probably has little or no use in most breeding programs. In contrast, a "single major fault" yellow card animal may well be beneficial in a herd in which the fault of this one animal is matched by strength in the other animals.

No single animal is best for all breeding programs.

A compromise between card grading and competitive showing has developed at some shows. The animals are card graded first, and then all of the blue card animals are ushered into the ring for a more traditional competitive show among those superior animals. This tends to satisfy the breeders' desire for a competitive show with one first place animal, while at the same time giving the nod to the strengths of the card grading system. The idea that no single animal is best for all breeding programs can still be communicated to the audience.

Non-competitive Exhibition

Exhibition, in contrast to showing, may well not involve any competitive aspect but may instead emphasize contact with the public and opportunity for educational exchange of information. Many opportunities for exhibition and education are similar to those for competitive showing as these can occur at fairs and stock shows, as well as at nature centers, museums, historic sites, schools and other public education venues. Private breeder members of associations frequently organize educational materials such as photo albums and breed literature stall-side in the exhibition barns. To these useful pieces of information can easily be added association literature that can be given to those visiting the exhibit and expressing interest in the breed.

The breed association for Leicester Longwool sheep in the USA has taken the stance that competitive showing is a potential risk for the breed, and has declined allowing sheep to be shown. The association encourages card grading and non-competitive exhibition as alternatives. Photo by D. P. Sponenberg.

Several breed associations have banned competitive showing and rely instead on exhibitions to disseminate information about the production characteristics of their breed rather than the visual characteristics. This strategy has been adopted specifically to minimize the potentially damaging aspects of competitive showing.

8. Registry

Functions of a breed registry are subtly different than those of a breed association, although most breed associations usually take on the functions of a registry. Strictly speaking, registry functions include all of the specific activities that revolve around documenting, managing and monitoring the mating and genetic contribution of individual animals of the breed, as well as the breed as a whole. Registry functions include a great many things that can help the genetic management of all breeds, and especially rare breeds.

Registration

The most basic function of a registry is to validate individual animals (or more rarely, groups of animals) as being of a specific breed. This is usually based on recorded pedigrees. Registration function is essential if a breed is to persist into the future. Registration assures breeders, regardless of where they are, that animals with the validation of registration are *bona fide* members of the breed. That knowledge allows breeders to confidently include such animals in purebred mating programs.

Validation of animals as representative of a specific breed is usually done by registration of individual animals. Registration links an individual animal by some sort of unique identification (description, tattoo, ear tag, microchip, brand, photograph) to a certificate. Individual validation is the most common approach with large animals of relatively high individual monetary value such as horses, cattle, donkeys, and alpacas.

Registry functions can help genetic management of breeds.

A different strategy has been used for poultry breeds. In general, poultry have no form of identity validation aside from external phenotype. As traditional breeds of poultry are gaining market acceptance as a premium product, a move toward flock validation is gaining momentum. Such validations could be accomplished by breed clubs, or by umbrella organizations such as the American Poultry Association. The goal is to validate populations of purebred poultry as members of their breed, which will aid breeders in marketing efforts. This serves some of the same functions as individual registration does in some of the larger mammal

Poultry breeds, like these Dominiques, have an increasing need for validation of breed status as the demand for certain products from certain breeds begins to increase. Photo by ALBC.

species, but avoids individual identification of birds, which tends to be costly and cumbersome relative to the economic value of individual birds.

In between the extremes of poultry and large stock lie swine, sheep, and goats. Nearly all breeds of these species do have individual registration and validation. Most swine breeds require litter registration, with individual registration of older or adult animals intended for breeding. For some breeds of sheep and goats, especially the adapted landraces, the individual–based approach may need to be reconsidered because herd-based approaches may be more practical. The challenge with a herd-based system is to develop an evaluation process that will assure breed integrity by only including purebred animals, while also avoiding the labor, effort, and expense that individual registration entails.

Pedigrees

Pedigrees are records of animal ancestry, and recording pedigrees is a function of most breed registries. The recording of pedigrees accomplishes distinct goals. Usually pedigrees validate an individual animal's ancestry within the breed. The validation of pedigree adds to the individual identity of an animal, and can help guide breeding choices, because the genetic background of an animal is more fully known than if pedigree information is lacking. Not only is this information useful in making breeding decisions, but it is also useful to buyers selecting animals to complement their own existing bloodlines.

Different registries trace ancestries back to different generational levels on pedigree certificates, but nearly all should include information back to grandparents. Pedigree information can be minimal and include only ancestor identification, or can be more detailed and include information about ancestors such as color, size, production characteristics, or show ring wins or other performance.

It is important to separate the concept of registration from that of pedigree records. Registrations specifically validate individual animals as being of a designated breed, whereas pedigree records specifically validate ancestry. Often the two are linked together so that a single certificate accomplishes both purposes. For some breeds only registration without individual pedigree documentation may be appropriate. This is especially true of landraces that experience multi-sire mating. Pedigree information for individual animals tends to be less and less available as animal size and individual value diminish. Individual pedigrees (and indeed registrations) are rare for poultry but routine for cattle and horses. Sheep and goats vary from breed to breed as to their requirements for registration and pedigree records, although the general trend is for pedigrees to be complete for these species. Within all classes of livestock, including poultry, individual breeders of elite stock do indeed document and rely on pedigree information. Registry involvement takes this activity to a level higher than the individual breeder so that the information becomes validated as animals move from breeder to breeder or owner to owner. Registration serves to make the information more publicly available and also to certify the accuracy of the information by a neutral

Registrations validate animals as being of a specific breed, whereas pedigrees specifically validate ancestry.

Spanish goats in the USA do not yet have a registry, and face the challenge of lack of pedigree information in populations managed under extensive conditions. Photo by D. P. Sponenberg.

party (registry) rather than one with a direct interest in the animal (breeder).

Registration papers indicate only that an animal is of a stated breed. Most registration certificates include an individual number that serves to track that animal and identify it and its pedigree, and most also include a printed pedigree. This is the reason for many people confusing registrations and pedigrees. The registration function is critical to the survival of pure breeds of livestock, while the pedigree is valuable for the genetic management of the breed, although each serves a different role.

Obtaining Pedigree Information

The validity of pedigrees varies immensely from breed to breed and species to species. Pedigree accuracy can very much become a sort of "holy grail" to breeders, but its usefulness must be put into a practical context or the pursuit of accuracy can easily overwhelm scant resources.

Validity of pedigrees varies breed to breed.

For species with high individual value (horses, cattle, alpacas) it is not too unusual for each animal presented for registration to be bloodtyped or DNA typed. This provides for parentage verification, which validates the pedigree that is stated on the registration application. The cost of such verification is a relatively small fraction of the value of the animal, and it makes sense in such situations to pursue this avenue of accuracy. For breeds raised in more extensive settings (range cattle, range horses) the practicality of this strategy begins to diminish because obtaining the required blood, tissues, or hair can be expensive and logistically challenging because animals need to be gathered, restrained, identified, and sampled.

Species with relatively low individual commercial value such as sheep, goats, and poultry, are almost never validated by checking the DNA of offspring and parents. The registries of these species rely on the word of the breeder about the parentage of the animals presented for registration. Some mistakes are inevitable in such a situation, some from intentional fraud but even more from honest mistakes that are a part of everyday life. Their significance depends on their extent. A few of the larger registries for these species undertake "spot checks" in which occasional animals are pedigree-validated by DNA test of the animal and its parents. This serves as a check to assure that misidentification is rare, and sends a signal to the potentially fraudulent breeder that they risk detection.

Pedigrees and Genetic Diversity

Pedigrees serve as an overall assessment of the genetic diversity within a breed. This can become critically important to the survival of the breed, because when genetic diversity becomes too restricted the overall vitality can begin to fail. This is occurring even in numerous breeds such as the Holstein cattle and Thoroughbred horses due to narrowing of the genetic base by using a high proportion of a few related individuals to produce the next generation. Pedigree analysis can quickly document the extent to which this is occurring, and can also lead to breeding strategies to help alleviate the problem by targeting the increased use of underrepresented lines.

The generation back to which pedigree information is useful is controversial. Information back to grandparents is truly useful in most decisions, but further back it becomes much less of a practical issue because the contribution of a great grandparent is small enough to be negligible. Some breeders make pedigree research a passion, and trace animals back through tens of generations. This is interesting historically, but has little biological impact in most situations. An animal's parents and grandparents are much more informative than any generation further back.

Pedigree Recording Systems

Pedigree information is usually considered to be of utmost concern to breed associations as one of the main reasons for interest in associations. Pedigrees have become synonymous with purebred livestock, and while their importance to certain breeds is less than others (landraces and poultry, for example), details of pedigrees can be essential in managing the genetic status of a breed.

Older methods to track pedigrees were nearly all static systems that involved either hand-written or typed lists. These served well for years, but have the drawback that their preparation can be very tedious for the registrar. Pedigrees are generally full of numbers (such as registration numbers) and also full of generally odd names including farm names and individual animal names. The character of the information is such that transcription errors are likely. While most errors do not result in any major disruption of the breed, they can greatly limit the ability to accurately probe into the genetic heritage and history of a breed.

New systems of tracking pedigrees include several computerized database systems. These are important tools in population management used by many different registry organizations. Nearly all large registries and umbrella organizations that serve several breeds use electronic database systems. Systems have been developed by different organizations that are very specific for individual breeds. Other more generic options are also available that will work with a wide number of breeds. Database systems generally provide for entry of the individual animal data such as registration number, sex, color, parentage, owner, breeder, and other important details. Then the program automati-

Computerized pedigree systems are useful tools

cally fills in the pedigree portion of the record from the sire and dam information. This approach reduces the chance of transcription errors because the same data do not require repeated manual entry into the system.

The value of the database systems is that they can be used to generate other important information such as inbreeding coefficients, get of sire or dam, animals under a specific ownership, geographic distribution in the face of a disease threat, or numbers of registrations per breeder.

A few widely used database programs include "Breed Society Record and Ped eView" by Grassroots Systems LTD, Gidleys, 36 Ide Lane, Alphington, Exeter, EX28UT, United Kingdom. Their email address is: sales@grassroots.co.uk. Another very successful one has been Breeders Assistant, http://www.tenset.co.uk/.

The International Species Identification System (ISIS) is a non-profit organization that works with, but is not exclusively for, zoological institutions worldwide. ISIS provides software and technical support for animal information databases, studbooks, and population management tools that are used to accomplish the long–term conservation management goals of their members. They developed the program *Single Population Analysis & Records Keeping System* (SPARKS). This software supports studbook management and species analysis.

Database systems for pedigree management can provide powerful analysis of breed dynamics.

SPARKS is DOS-based software used by hundreds of studbook keepers worldwide. The program is regularly updated and has evolved over the years to become increasingly user friendly and versatile. It is possible to import files from Microsoft EXCEL directly into the SPARKS program. SPARKS offers the option to manage populations by groups if individual identification of animals is not possible, as is the case with much of the poultry and some landrace breeds kept in the USA. There is a secondary software program, PM2000, that is used as a population management tool in conjunction with SPARKS. PM2000 can be used as a stand-alone program for population management, if the demographic and genetic data are first prepared in a specific format. Population managers use both programs to make breeding decisions and manage populations of animals in captivity. (For more information, see http://isis.org/CMSHOME.)

Litter Recording

Swine, dog, and rabbit breeders are likely to use litter recording systems because these serve well to track the multiple offspring born to these species. Specific protocols for litter registration vary, but generally the breeder of a purebred litter notifies the association of the birth of the litter, with information about parent-

Litter recording is important in maintaining information on swine breeding activity, such as for these Ossabaw hogs. Photo by D. P. Sponenberg.

age of the litter, numbers, and sexes of the offspring. The litter registration is frequently followed up later with individual registration of selected animals in the litter. The recording of the litter simply makes the members of the litter eligible for later individual registration. This serves as an important check on accuracy of registration information for litter-producing species.

Stud Reports

Stud reports are used by some breed associations, usually for horse breeds, to track the level of breeding activity that is accomplished within the breed. These are usually annual reports that are filed by breeders after the breeding season. They generally include identification and registration information on the male, and then similar information for all females to which he was mated during the year's activity. These are used to anticipate the production of the following year. It is common for breed associations that use stud reports to restrict subsequent registration only to those animals whose parents are included in stud reports. These are used to monitor breeding activity, and also as a check on the accuracy of the identity information of animals presented for registration.

Selective Recording Systems

Breeders generally register only those animals that they are likely to use themselves or to sell as purebred breeding stock to other breeders. Rabbits must be registered to compete in sanctioned shows, so only a small percentage of the purebred animals are documented in the registry system. Non-breeding animals of any breed or species are unlikely to be registered (with the exception of horses and dogs). As a result, registrations are an incomplete record of purebred production. The incompleteness is a consequence of decisions by the breeders. This results in little practical loss of information, because eliminating non-reproducing animals more accurately reveals the level of pure breeding within the breed than would be reflected by registration of every animal produced. Incomplete registration of purebreds indicates that culling is occurring, which means that breeders are being selective. This is generally good for any pure breed, because not every purebred offspring offers genetic excellence to the breed.

Incomplete registration can indicate that culling occurs.

Some breed associations have restrictions on registration, so that decisions regarding which specific animals to register are imposed by the association and not by the individual breeders. Few such associations exist in the USA, but Warmblood horse breeding in most of Europe is characterized by this system. Breeding stock, and especially males, must be evaluated by a committee for conformation and performance before being allowed to stand as a registered breeding animal. These tightly controlled systems can enhance and improve a breed, and in some situations are essential to conserving the breed as a traditional genetic resource. Selective registration, however, can succumb to fads or current fashions, resulting in a change within the breed from a traditional to a more modern and uniform type. This is especially the case with Warmblood horse breeding where the blending of bloodlines and the selection to a single ideal has assured homogenization of these once unique breeds.

Clun Forest sheep breeders use a different and useful strategy for selective registration of lamb crops. In this breed it is permissable for breeders to register only a percentage of each lamb crop. The specific animals included in the registration are left to a breeder's discretion, but the rule that only a percentage of lambs born can be registered forces each breeder to exercise selection in the lamb crop. This approach does not constrain the specific strategy or philosophy for keep-or-cull decisions, but does insist that these decisions must be occurring in every flock. A potential drawback is that this, and all registration systems, ultimately depends on the honesty of the breeder in filling out total lamb production so that the correct number can be ultimately registered.

Registrations Are Important

Registration figures can give an accurate picture of purebred breeding within a breed, and can do so much more quickly than other measures such as a complete census of live animals. As an example, any breed that is commonly used for crossbreeding (such as Finnsheep or American Brahman cattle) may have a much higher census than would be reflected in the registration figures. As a result, any assessment based on total census would significantly overestimate the level of purebred breeding that is occurring.

Registrations allow for tracking individuals within a breed. Regardless of how many times a specific animal changes hands, the registration certificate can follow in order to maintain that animal's individual identity within the breed. By documenting the identity of animals it is less likely that the animals will be lost to the breed, especially in cases such as death of owner or other disruptive catastrophes.

Registrations can also help to identify and salvage genetic material that might otherwise be lost to a breed. A perplexing minority of breeders fail to register purebred stock, and then at a future date these breeders will appeal to breed organizations to register animals after a lapse of a few generations of registration. In some cases this can rise to the level of genetic blackmail. The rationale for including these lapsed but purebred animals back into registered stock is compelling on the biological level, but on the political level is unsavory because it rewards irresponsible behavior on the part of those who let registrations lapse. While legitimate reasons do exist for non-registration, in most cases the sad truth is that owners and breeders fail to register animals through laziness, selfishness, or false economy.

Registrations help breeds to monitor the status of breed populations and dy-

The American Brahman is widely used for crossbreeding, so that using census figures alone may overestimate the level of purebred breeding that is occurring. Photo by D. P. Sponenberg.

namics of the different lines and strains within the breed. Much of this monitoring is only possible if a reasonably high percentage of the breed is registered. In the absence of a high registration percentage it is nearly impossible to gain an accurate idea of breed populations or trends.

Closed Herd Book Registries

Many associations have "closed" registry herdbooks. This refers to a requirement that only animals with a registered sire and dam are themselves eligible for registration. No progeny that originate from unregistered stock are considered for registration. This requirement serves to make the population completely closed to outside genetics. This is the most common strategy for standardized breeds, and is especially true of international breeds of British origin. This practice of complete genetic isolation of breeds arose in the 1700s or 1800s from English work with the Shorthorn cattle and Thoroughbred horse herdbooks, which were organized around this principle. By the late 1800s and early 1900s this strategy had become synonymous with the term "purebred livestock," so that even landrace breed associations such as that for Texas Longhorn cattle eventually adopted this same defining principle.

Closed populations do serve to isolate breeds, and as a consequence serve to consolidate them as repeatable, predictable genetic packages. This is particularly useful in the early history of breeds. In mature breeds this strategy can be very constrictive as genetic variability (essential for vitality) is lost in each succeeding generation of breeding within a closed population. Because some of the earliest closed breeds are now in their second or third century, strategies for

Shorthorn cattle were among the first of all breeds to have a registry. Photo by D. P. Sponenberg.

managing pure breeds as repeatable genetic packages while retaining enough variability for vitality promise to become increasingly important.

For a closed herdbook to be valid in a genetic sense, it should include a vast majority of the breed. It is all too common for recently structured landrace herdbooks to close before a reasonably high proportion of the animals that are actually members of the landrace have been included. Closing the books restricts the numbers of breeders and animals within the population recognized as valid by the registry; the included breeders benefit from the restricted supply. This is shortsighted when considering the importance of the fate of breeds as genetic resources. With few exceptions closed herdbooks are inappropriate for landrace or local breeds.

Open Herd Book Registries

Few purebred breed associations follow an open herd book principle for organizing the breed population. This strategy is most common among landrace populations for which it is likely that many good representatives of the population remain outside the registered population. That is, the registered populations of these breeds tend to represent only a portion, and frequently a small portion, of the entire breed. In such breeds it is wisest to keep the herdbook open to allow for inclusion of these purebred animals as they come to the attention of the association.

Herdbooks for most landraces should remain open so that new candidates can be included and the entire breed can be recognized.

The key to operating an open herd book that works as an effective tool for genetic resource conservation is to assure that only representative animals are brought in. The goal should be to include every legitimate representative of the breed, and to exclude all non-representative animals. This is a tough line to draw, and will vary from situation to situation and breed to breed. A few examples may help.

A group of traditionally-minded Texas Longhorn cattle breeders banded together after the main portion of the association began going down a path selecting for big, smooth, longhorned, speckled cattle that diverged dramatically from the original type. The original association has a closed herdbook, but it was closed after the inclusion of some few cattle with Hereford and presumed African bloodtypes. The traditionally-minded breeders then decided to develop an association that would continue to accept cattle after a visual inspection, and following success at that step, bloodtyping to assure that only Iberian bloodtypes of the traditional Texas Longhorns were present. Offspring of registered cattle could then be registered without further inspection or bloodtyping, so that only the newly accepted foundation animals needed to go through the inspection and genetic documentation step. The result has been that the new association, though maintaining an open herd book, has been able to more closely guard the pure genetic resource than have older associations with prematurely closed herdbooks.

In contrast, the Navajo-Churro sheep breeders have chosen a system that inspects each and every animal presented for registration, even if it comes from a long line of registered sheep. The goal is to assure that the traditional phenotype is recognized and perpetuated. This includes fleece type and structure, which is all-important to this breed. This more restrictive approach assures that the desired phenotype and its underlying genotype does not drift out of existence. This is an excellent example of selection to type assured by the requirements of the association, and relates back to the functions of a successful association of breeders in crafting a breed as a genetic resource. This procedure can be resource-intensive, but is managed so as to be practical. Breeders in areas where the sheep are numerous can get their sheep inspected on-site. In addition it is possible to submit photos and fleece samples, a strategy that allows more far-flung isolated breeders to participate in the breed. Recent designation of some Navajo-speaking inspectors has greatly facilitated the participation of this important traditional community in the conservation of this important sheep breed.

Several American landrace populations use open herdbooks to good advantage. Among these is the large group of Colonial Spanish Horse registries. Many of these will accept new horses following a

Navajo-Churro sheep have greatly benefited from their open flock book with strict inspection procedures. These have secured this breed a sound future as a useful genetic resource. Photo by Don E. Bixby.

Tennessee Myotonic (Fainting) goats have open registries that try to incorporate unregistered herds of this landrace. Photo by D. P. Sponenberg.

visual inspection and an evaluation of the history of the horse and its population of origin. The associations vary in the extent to which the books function as open books – some add horses relatively easily, while others are much more restrictive. Either philosophy presents both positive and negative aspects, depending on the overall goal for conservation purposes.

Florida Cracker and Pineywoods cattle breed associations both use an open herdbook. This works well when old lines of these breeds are newly discovered. As time proceeds, fewer and fewer of these are discovered, but they do still come to light occasionally. Any such discoveries are genetically important to these breeds, and need to be fully included in the breed.

The Tennessee Myotonic (Fainting) goat registries are also open. The International Fainting Goat Association requires photographic documentation of the stiffness of the goats presented for registry. This presents a subtle problem in that not all myotonic (stiff) goats are of traditional Tennessee breeding, so that the breed has become confused with this single genetic trait. As a result, it is likely that nontraditional goats are registered along with traditional goats, and this poses a threat to the original genetic resource. This situation has no easy solution, especially because many traditional Tennessee Myotonic Goats remain unregistered and to close the books would be to forever deny them their legitimate role in the breed's future. Recent developments in the Myotonic Goat Registry are attempting to surmount these problems.

Open herdbooks are appropriate for many breeds, especially American landraces.

Open herdbooks are entirely appropriate for many breeds – especially for those with local American origins. Such a strategy for breed maintenance, though, brings with it great responsibility for the association to be clear-headed about the character and importance of breeds. As breeds go in and out of favor it can easily happen that during periods of popularity many non-representative animals are presented for registration. And, in periods of less interest it is easily possible for good, representative animals to slip from the registered population and potentially be lost from the breed. Either situation is bad for the breed, but no easy solution solves both problems.

Registration of Crossbreds and Partbreds

Over the years the registration or recording of crossbred animals into separate sections of purebred registries for a number of breeds has been a very contentious issue. On one side of the issue are people who are adamant that no crossbreeding should ever occur with the breed, and that to do so is to assure the demise of the breed. On the other side are breeders who hold that any and every animal with a drop of purebred blood should be registered.

Recording or registration of crossbreds gets back to the genetic definition of breeds, and the need to maintain them as genetic pools. In addition, it gives credence to the role that several breeds have to play in crossbreeding for commercial excellence. Maintaining the consistency of the entire package should be of utmost interest to purebred breeders. To lose the package is to lose the breed.

In order to protect the genetic integrity of several breeds it is prudent to identify and register known crossbreds. This allows breeders to "grade up" to purebred status, by clearly identifying animals that are known to be partly outside breeding. By this strategy, even if a 75% purebred animal looked purebred it could not be misidentified and misregistered as a purebred animal. Registering the crossbreds or upgrades in herds of serious breeders gives these animals a specific identity. This greatly reduces the chance that they or their offspring will be presented to the registry as purebreds (whether deliberately or accidently). In order to prevent the registration of crossbreds as purebreds, it may be wise to provide a specific strategy for the registration of crossbreds and upgraded animals. Several registries have developed such strategies.

In addition, registration of crossbreds can eliminate any temptation to register them as purebreds based on performance or phenotype. For example, some pony breeds have a modern "sport" pony type in addition to the more traditional cart pony or general farm use type. The traditional type has been essential in providing certain characteristics to the crossbred sport type, but is in danger of disappearing if the modern, crossbred type is allowed to be registered into the purebred herdbook. Registration of crossbreds in an appropriate section of the herdbook allows them to reflect the excellence of the pure breed, and to serve as ambassadors for the role of the pure breed in producing this sport type.

9. Monitoring Breed Populations

Monitoring the census of a breed is an important registry function. Censuses are especially important for rare breeds because of the considerable risk of rapid and serious numeric depletion or loss of critical bloodlines. Trends must be monitored as an ongoing process to avoid being surprised by sudden, unanticipated changes. Associations monitor the census in different ways.

One relatively easy way to monitor the census is to simply monitor annual trends of registrations. This is an easy and useful measure of purebred recruitment, and has numerous advantages over other parameters as a single measure of a breed's numerical health. The number of new registrations counts those very animals that are most likely to contribute to the next purebred generations.

Registration figures have a significant advantage over other measures such as total number of breeding animals. This is especially true for breeds that are largely used for crossbreeding, because many registered animals may not be contributing to purebred replacement. Counting only adult breeding animals in such breeds will overestimate the level of purebred breeding. Purebred replacement of horse breeds is also likely to be vastly overestimated by counting only adult animals capable of reproduction (stallions and mares) because many mares are deliberately kept from reproduction and are only used for athletic performance or as companion animals. Registration figures neatly and efficiently get around this problem.

Censuses are important for rare breeds.

For many rare breeds a simple count of annual registrations is not sufficient to monitor the genetic status of a breed. More detailed analyses are important for these breeds, including the numbers of animals within the various distinct bloodlines or families of the breed. This can be an arduous task, but computerization of registry records can be a great help. If bloodlines are linked to pedigree information, then automated analyses are possible and tedious hand-done analyses can be avoided.

A consistent problem with advanced analysis of a breed census is that many factors within a registry tend to overestimate the living population of registered

Mammoth Jacks, and other breeds widely used for crossing, are more accurately assessed by registrations than by actual census. Photo by Mary Ellen Nicholas.

animals, let alone the entire population of the breed whether registered or not. Breeders usually do not relay information to the association concerning animals that are removed from the population by death or sale to nonpurebreeding situations, so relying on registry records for a total count is generally inaccurate.

An opposite problem is that registered populations always underestimate the actual population for many landrace populations to a far greater extent than for standardized breeds. The significance of this varies from landrace to landrace. For landrace associations with reasonably open herdbooks this can be a large problem as it becomes nearly impossible to get an accurate estimate of breed populations. For those with essentially closed herdbooks the issue diminishes for the registered population, because entering new individuals of unregistered background into the registered population is difficult or impossible. For those breeds, though, the status of the breed as a registered population versus a functional agricultural entity represented by purebred but unregistered animals can be important. Exceptions where standardized breeds are underreported are important, and include some dairy cattle situations as well as breeder sows where registration documentation is seldom used. If the registered population excludes significant numbers of purebred animals, then the association is not fully accomplishing its mission to monitor and conserve the breed.

While census figures can rely on annual registrations in registered populations, in unregistered populations this is obviously impossible. Poultry, specifically, are not registered by any association, and keeping track of numerical trends in poultry breeds is very difficult. Surveys of a few key breeders and seasonal hatcheries can be very helpful. Surveys, though, are notoriously easy to disregard and many fail to be returned. Telephone interviews are frequently required to get information on population numbers of poultry breeds. Poultry present another challenge because many small but potentially important populations are likely to be overlooked by any census technique. The inevitable tendency for poultry census work to focus mainly on large hatcheries or very engaged private breeders does have the advantage of targeting the very segment of the breeders that is likely to reflect the breed's status. As a practical issue, poultry census work does establish the "worst case scenario" reasonably well by consistently underestimating numbers.

It is important to know both numbers and which populations are being bred to the standard.

Poultry present another challenge because it is difficult to sort through the relative importance and quality of breeding groups. Not only is the numeric assessment important, it is also important to have some idea of which populations are being bred to the standard, which are being selected for production characteristics, and which are purebred from a base of the original founding genetics. The more qualitative aspects are very difficult to assess by any other than a site visit, and this is impossible in all but a very few situations. As a result, the census work for most poultry is difficult and has inherent inaccuracies. While details may be less finely focused than those for livestock, it still remains true that census figures, whatever their drawbacks and inaccuracies, give a very

Breeds of poultry, such as these Saxony ducks, are notoriously difficult to census because they are not registered. Photo by D. P. Sponenberg.

good idea of the relative status of breeds. They are especially good at pointing to those breeds in critical need of help.

Rabbit breed registration issues lie somewhere between larger farm animals and poultry. Registration is required for any animals entered into competitive showing. This ensures that show animals will be registered, but gives no clear indication of the number of either the reproducing population or the total population.

Monitoring Breed Health

Monitoring breed health has multiple aspects. One of these is the general genetic status of the breed. This involves inbreeding, availability of sub-lines, and other related issues. These are complicated issues, and are all addressed in separate sections due to their complex but important character. A second aspect is specific genetic-related diseases or traits that might be present in a breed. A third aspect is general health and vitality of the members of the breed and how that affects the breed, its breeding, and its use.

Congenital and Genetic Defects

Congenital defects include any defect present at birth. This is an important concept, because the definition does not imply any genetic cause. While some congenital defects are genetic in origin, many are developmental and may include several viral- or environmentally-induced defects. Conversely, some genetic defects are not noticeable at birth but only become evident later. Examples include some metabolic storage diseases, or the recently important bovine leukocyte adhesion defect (blad) which only surfaces later in life.

Congenital defects are not all genetic.

It is wise for breed associations to track both currently known congenital defects as well as newly emerging ones. This forward-looking approach allows associations to quickly identify and develop a strategy for managing all defects.

Developing a reporting method for defects is no easy task. The most successful methods are those that are employed by the most populous dairy cattle breeds. These provide for a description of the defect as well as for pedigree information. The reason these work so well in dairy cattle breeds is that calves are nearly always conceived by artificial insemination so owners of dams have no vested interest in protecting the sire's reputation, but instead have every incentive to quickly uncover any genetic weaknesses that may be present.

The situation in most other breeds is opposite that of dairy cattle breeds, because breeders of other breeds generally have a direct economic stake in the reputation of the sires involved in producing defective offspring. Silence is a more likely response than full disclosure of faults. No strategy can completely eliminate this incentive, but a few tactics can help reduce it. One useful option is to make reporting anonymous and not linked with pedigree information. The benefit of this is that the association can then keep tabs on the overall incidence rates of various defects. If a specific defect begins to increase, then steps can be taken to educate breeders, investigate the cause of the defect, and potentially develop or employ a test for carriers if the defect is proven genetic.

Levelheaded and fair policies towards genetic defects are tough to formulate, but are essential if success is to be achieved in discovering defects and reducing their rates of occurrence. For populous breeds fairly strict measures are appropriate, such as elimination of all carriers, or requiring that males be certified free of a given defect that might be common in the breed. For rare breeds this approach can be problematic, because the elimination of breeding stock takes away the entire genome of carriers and not just the offending gene. For example, over-zealousness can eliminate complete breeding lines of horses that meet the breed standard but fail the gene carrier test for a coat color gene.

Policies for defects must be fair and wise.

For other traits, such as junctional epidermolysis bullosa, a lethal skin defect of Belgians and American Cream horses, it becomes a little more difficult to make simple decisions. This defect is recessive, and is lethal to foals. Carriers

American Cream Draft horses have benefited from forward-looking testing and reporting regimens for the junctional epidermolysis bullosa defect that is lethal to foals. Photo by Karen Smith.

can be identified by blood test, but if all carriers are eliminated from reproduction, then a rare breed such as the American Cream moves ever so much closer to extinction by loss of breeding animals and genetic variability. An alternative strategy is to insist that all stallions be tested and the results be made public. Carrier stallions can (and should) be mated to non-carrier mares to produce replacements, with the eventual goal of generating sons that are not carriers so that the bloodlines are not lost but the offending single gene is. This can take several generations to accomplish, but breeders should all be encouraged to move toward that goal while preserving the genetic integrity of the breed by not discarding carriers too hastily.

Carriers should be mated to noncarriers.

Discovering which defects are present in rare breeds can be a great challenge. Most published accounts of genetic defects appear in veterinary or genetics journals, which are unlikely to be widely available to the general public because their cost prohibits all but large university and research libraries from subscribing. In addition, once the defects are described in any one breed they are unlikely to again appear in publications because reports that only document the occurrence in an additional new breed are unlikely to be accepted for publication. This phenomenon results in the literature tending to underreport the breed incidences of defects. The best strategy to circumvent this problem is for breed associations to be forward-looking and to carefully document and then make available information on the various genetic and other defects that have been found in the breed.

10. Breeder Responsibilities

Breeders have a great many responsibilities to the breeds that they manage, which is a very healthy and a completely appropriate state of affairs. Breeders gave us the breeds we cherish, and breeders are certainly capable of managing them to pass on to future generations. Without engaged and dedicated breeders, breeds lose their relevance in the agricultural landscape and risk being relegated to the status of trivial artifacts or face extinction.

It is the responsibility of breeders to manage the breed type to fall within the breed range. This implies that both management and selection will be appropriate to the specific breed genetic resource that is being managed. This can be a subtle and powerful concept, and is likely to be overlooked by many breeders. While many breeds will thrive under ideal management and abundant resources, some breeds will only maintain their traditional genetic heritage by being placed in environments that challenge them to retain their adaptive traits and remain productive. It is the breeders of the breed that must provide for that environment, and for the selection of the animals that are best adapted to it.

Breeders gave us breeds, breeders can manage them for the future.

Breeds are an important part of food security. The loss of genomes, either through breed extinction or through breed changes, reduces the options for future food production strategies. Industrial strains are proving to have very reduced biological fitness, and may well not be able to adapt outside of their narrow, if exquisitely productive, agricultural setting. Different breed choices must be available, intact, and viable in order for the demands of changing agricultural systems to be met.

Breeders also have important responsibilities to their associations and registries. Among these are providing accurate and timely information. This can include notification of births, deaths, and ownership transfers, all of which help the registry to maintain accurate and current records. More contentious but equally important is alerting the association to the production of any defects or known genetic diseases within pure or crossbred examples of the breed. Only by having this information can associations take early and effective measures to assure that the breed remains healthy and viable.

Breeds, Breeders, Associations, and the Future

The future has always seemed dark and mysterious, and doomsayers have always pointed to a downward spiral of culture and life throughout all eras of history. Without being unnecessarily pessimistic, though, it is possible to point to some very real threats to breeds, breed integrity, the function of agricultural systems, and what the future might hold. Fortunately, it is also possible to point to some bright spots on the horizon – bright spots that are presently increasing in brilliance as well as size.

Different breed choices must be available for changing agricultural demands.

Several threats that are unique to this time in history confront breeds and their integrity and use in agricultural systems. Some threats are subtle and internal. Among those is the philosophy of absolute breed purity that keeps breeds completely isolated genetically from all outside influences. This model for breeds and their maintenance is rather recent, having developed only in the last couple of centuries. This model differs from the traditional course of breed development, which insisted on utility and predictability but only as they served functional ends. Anything that contributed to the predictable package was considered fair game, and the concept of complete

It is important for every breed association to enlist and encourage the next generation of breeders. These girls are enjoying their Jacob lambs. Photo by Bob May.

Maintaining some breed uses, such as Pineywoods oxen, is a huge challenge in today's culture where this service has been largely replaced by machines. Photo by D. P. Sponenberg.

genetic isolation was not in force. It remains to be seen whether complete genetic isolation will eventually result in a gradual, ever-tightening constraint of inbreeding depression for many breeds. Of all the threats facing breeds, this may be the most insidious and dangerous, because it is imposed by breed advocates that are in no way trying to diminish breeds and their utility.

Communication and transportation advances in the last century have also posed a threat to breeds and their integrity. This threat has usually arrived from a gradual homogenization of regional and international cultures so that unique products and the animals that provide them are much less valued than once they were. Communication and transportation have resulted in ever increasing consolidation of the production and marketing of agricultural products so that point-of-origin producers must follow the dictates of this consolidation to be successful. Globalization is the final stage of this process, and if unchecked can result in a very severe diminishment of genomes worldwide which produce unique, satisfying, healthy, and interesting local products.

Increasingly the cultural environment for breeds and breed maintenance has also changed. Breeds which were once valued as essential ingredients to local and regional agricultural production and cultural identity have become somewhat trivialized as lifestyle endeavors for those wealthy enough to indulge themselves in this activity. Breeds, while saved, have moved from essential partners in survival to a nonessential pet or hobby status. This switch in cultural environment changes the selection environment in which the breeds survive and persist, and can only result in genetic changes as well.

Not all is doom and gloom, though. An increasing number of people, both producers and consumers, are realizing that a sustainable and local agricultural system has great advantages for people, animals and the environment. The growth of this view of agriculture will help to provide rare and traditional breeds a secure future as the connection of breed, place, and production system becomes recognized and appreciated by larger numbers of people.

The future of breeds lies with breeders.

Finally, it is important to remember that generations of breeders have given us the breeds that we enjoy and use today. The future hope of breeds and their conservation lies with breeders as stewards. With a few tools and some encouragement they are very much equal to accomplishing this important task and accomplishing it successfully.

Appendices

Appendix 1. Colonial Spanish Horse Score Sheet

The following is a Colonial Spanish Horse score sheet developed by D. P. Sponenberg and Chuck Reed. Horses are scored on various aspects of conformation and type. The final result is not a simple average of scores, but rather a close look at the number of not typical (high) versus typical (low) scores. The head character weighs in heavily in the final determination, especially if the body scores well. Put another way, a high-scoring body with a low-scoring head is still rejected because these horses are unlikely to be Colonial Spanish.

MOST TYPICAL – SCORE 1	NOT TYPICAL – SCORE 5
HEAD PROFILE	
Either: • concave/flat on forehead and then convex from top of nasal area to top of upper lip (subconvex). • uniformly slightly convex from poll to muzzle. • straight.	• dished as in Arabian. • markedly convex.
HEAD FROM FRONT VIEW	
Wide between eyes (cranial portion) but tapering and "chiseled" in nasal/facial portion. This is a very important indicator, and width between eyes with sculpted taper to fine muzzle is very typical.	Wide and fleshy throughout head from cranial portion to muzzle.
NOSTRILS	
Small, thin, and crescent-shaped. Flare larger when excited or exerting.	Large, round, and open at rest.
EARS	
Small to medium length, with distinctive notch or inward point at tips.	Long, straight, with no inward point at tip. Thick, wide, or boxy.

MOST TYPICAL – SCORE 1	NOT TYPICAL – SCORE 5
EYES	
Vary from large to small (pig eyes). Usually fairly high on head.	Large and bold, low on head.
MUZZLE PROFILE	
Refined, usually with the top lip longer than the bottom lip.	coarse and thick with lower lip loose, large, and projecting beyond upper lip.
MUZZLE FRONT VIEW	
Fine taper down face to nostrils, slight outward flare, and then inward delicate curve to small, fine muzzle that is narrower than region between nostrils.	Coarse and rounded, or heavy and somewhat square as the Quarter Horse, rather than having the tapering curves of the typical muzzle.
NECK	
Wide from side, sometimes ewe-necked, attached low on chest.	Thin, long, and set high on chest.
HEIGHT	
Usually 13.2 to 14.2 hands high. Horses over 15 hands are not typical.	Under 13 hands or over 15 hands is not typical.
WITHERS	
Pronounced and obvious. "sharp"	Low, thick, and meaty.
BACK	
Short, strong.	Long, weak, and plain.
CROUP PROFILE	
Angled from top to tail. Usually a 30 degree slope, some are steeper.	Flat or high.
TAIL SET	
Low, tail follows the croup angle so that tail "falls off" the croup.	High, tail up above the angle of the croup.
SHOULDER	
Should be long, and 45 to 55 degrees	Short, and steeper than 55 degrees
CHEST SIDE VIEW	
Deep, usually accounting for half of height	Shallow, less than half of height.
CHEST FRONT VIEW	
Narrow, and "pointed" in an "A" shape.	Broad, with chest flat across.
CHESTNUTS	
Small, frequently absent on rear, and flat rather than thick.	Large, and thick.
COLOR	
Any color. In populations the black-based colors are relatively common. No bonus points for any color, no suspicion of impurity on any color.	No color is penalized.

MOST TYPICAL – SCORE 1	NOT TYPICAL – SCORE 5
REAR LIMBS FROM REAR VIEW	
Straight along whole length, or inward to have close hocks and then straight to ground ("close hocks"), or slightly turned out from hocks to ground ("cow hocks") but not extreme. Legs very flexible. At trot the hind track often lands past the front track.	Excessive "cow hocks." Heavy, bunchy gaskin muscle, tight tendons.
FEATHERING ON LEGS	
Absent to light fetlock feathering, though some have long silky hair above ergot and a "comb" of curled hair up back of cannon. Some horses from mountain areas have more feathering than typical of others, and lose this after moving to other environments.	Coarse, abundant feathering as is seen in some draft horse breeds.
REAR	
Contour from top of croup to gaskin has a "break" in line at the point of the butt.	Contour from top of croup to gaskin is full and round "apple butt" with no break at the point of the butt.
HIP FROM REAR	
Spine higher than hip, resulting in "rafter" hip. Usually no crease from heavy muscling.	Thickly muscled with a distinct crease down the rear.
HIP FROM SIDE	
Long and sloping, well angled, and not heavy.	Short, poorly angled.
MUSCLING	
Long and tapered.	Short and thick "bunchy."
FRONT CANNON BONES	
Cross-section is round. Best to palpate this below the splint bones.	Cross section is flat across the rear of the bone.

Appendix II. Sample Bylaws

ARTICLE I
TITLE, OBJECTS, LOCATION

SECTION 1. TITLE The Association shall be known as ____, and shall at all times be operated and conducted as a non-profit corporation in accordance with the laws of the State of _____ for such organizations and by which it shall acquire all such rights as granted to associations of this kind.

SECTION 2. OBJECTS The purpose of the Association shall be to collect, record, and preserve the pedigrees of _____, to publish a breeding registry to be known as the _____Herd Book, and to stimulate and regulate any and all other matters such as may pertain to the history, breeding, exhibition, publicity, sale or improvements of this breed.

SECTION 3. PLACE OF BUSINESS The principal place of business for the ___ shall be in the state where the current registrar resides, but its members or officers may be residents of any state, territory or country, and business may be carried on at any place convenient to such members or officials, as may be participating.

ARTICLE II
MEMBERS

SECTION 1. MEMBERS

A. Membership. Membership is annual, and members are those people paying dues upon joining the Association and renewing their membership by paying dues in January of each year.

B. Members may be individuals, partnerships, or corporations.

C. As a condition of membership in the Association each member shall agree to conform to and abide by the Bylaws, Rules and Regulations of the Association, and amendments or modifications thereto, which may from time to time be adopted.

D. Application for membership may be made by submitting to the Registrar of the Association an application in the form prescribed by the Board of Directors, accompanied by the established membership fee.

E. All animals shall be registered under a single herd name unless they are owned by a partnership. In the case of a partnership a partnership agreement shall be placed on file with the current registrar of record at the time of the herds formation.

F. The Board of Directors shall have the power to accept or reject applications for membership, fix membership fees, and establish Rules and Regulations covering the rights and privileges of members, consistent with the provisions of these Bylaws.

G. Only Active members shall be entitled to vote on any matter submitted to a vote of the Membership. Each Active member shall have one vote. Corporations or firms who are Active Members shall designate in writing an individual officer, director or member of the corporation or firm, who shall exercise on behalf of the corporation or firm, the rights and privileges of such membership, including the right to vote and hold office. Husband, wife and children under the age of eighteen are entitled to only one vote, even if each owns animals in his own name. An exception to this rule would be in such cases where spouse or children have purchased separate memberships.

H. Membership in the Association shall cease upon the death, resignation or expulsion of a member, except as may otherwise be provided in the Rules and Regulations of the Association. Membership is not transferable.

I. The Board of Directors may provide for the issuance of Certificate evidencing membership in the Association.

SECTION 2. ANNUAL MEETINGS. The regular annual meeting of the members may be held at such time and place as may be fixed by the resolution of the Board of Directors for the purpose of electing directors and for the transaction of such other business as may be brought before the meeting.

If an annual meeting is to take place, notice of the meeting shall be given by mailing written notice stating the time and place of such meeting to each member's last known address as it appears on the Association's records not less than thirty (30) days prior to the date of such meeting.

If an annual meeting is not practical or feasible, members may address concerns, questions or other business via electronic means, regular mail or telephone to any Board Member to relay on to the rest of the board.

SECTION 3. SPECIAL MEETINGS. The President, or a majority of the Board of directors may call special meetings of the members, by giving written notice to the membership of the time and place of such meeting at least fifteen (15) days in advance. At a special meeting the members may transact only such business as is properly specified in the notice of meeting.

SECTION 4. QUORUM AND PROXY. For the purpose of election and the transaction of other business, the quorum shall consist of fifteen (15) or more voting members or ten percent (10%) of the total voting membership present in person, which ever shall be the least. Voting by proxy shall not be permitted.

ARTICLE III

DIRECTORS

SECTION 1. GOVERNMENT. A Board of Directors shall govern the business and property of the Association.

There shall be no less than three nor more than nine directors, as established from time to time by the Board or majority vote of the members at any annual or special meeting.

The initial Board of Directors shall consist of six directors, two of whom are to serve for three year terms; two of whom are to serve for two year terms; and, two of whom are to serve for one year terms with such terms to be determined by lot.

Directors elected to succeeding terms will be for a full three-year term. Directors must be Active Members of the Association.

SECTION 2. ELECTION. Elections for members of the Board of Directors may be conducted at the annual meeting or a vote by mail. Members may make nomination suggestions to the Nominating Committee for their review.

No less than thirty (30) days before the membership vote, the Board of Directors shall mail a list of the nominees and their resumes which will not exceed 200 words, to the members eligible to vote at that date. Memberships that have lapsed shall be sent notices of dues, which are due along with the ballot and resumes.

The CPA firm or attorney who will tally the ballots shall be chosen by the Board of Directors.

SECTION 3. GEOGRAPHIC DISTRIBUTION. Directors need not be citizens or residents of the United States of America.

It is declared to be the policy of the Association to have the various areas in which _____ are bred to be fairly represented on the Board of Directors, and it is hereby provided that no more than three Directors may be residents of the same state of the United States of America or residents of the same foreign country. In this instance, the term residence is defined as the state or foreign country in which the headquarters of a particular operation is located. The Headquarters of this Association may not be moved from the boundaries of the United States.

SECTION 4. VACANCY. If a Director, during his term of office, shall die or resign, or shall he disperse his herd and cease to be an active breeder, or shall fail to attend three consecutive meetings, or otherwise fail to perform the duties of a Director, the Board of Directors may, after appropriate notice to such Director, remove him from office and declare a vacancy. The Board of Directors may then fill the vacancy by appointment of a new Director for the unexpired portion of the term.

SECTION 5. RULES AND REGULATIONS. The Board of Directors shall have the power to establish Rules and Regulations for the conduct of the members of the Association and for the conduct of the affairs of the Association consistent with the provisions of these Bylaws.

SECTION 6. COMMITTEES. The Board of Directors may, from time to time appoint standing or special committees which may include nonmembers of the Board of Directors. Standing or special committees appointed by the Board of Directors shall be charged with and limited to such responsibilities as the Board of Directors shall set forth by resolution.

SECTION 7. ANNUAL AND REGULAR MEETINGS. The regular annual meeting of the Board of Directors may be held either in person, via conference call, or other electronic means and no notice shall be required for any such regular meeting of the Board. The Board, by rule, may provide for other regular meetings at stated times and places, of which no notice shall be required.

At an annual meeting, the Board shall proceed to the election of officers of the Association.

SECTION 8. SPECIAL MEETINGS. Special meetings of the Board of Directors shall be held whenever called by direction of the President or by two-thirds (2/3rds) of the Directors in office.

The Secretary shall give notice of each special meeting by mail or telephone to each Director at least ten (10) days before the meeting; but any Director may waive such notice. Unless otherwise indicated in the notice thereof, any and all business may be transacted at a special meeting.

SECTION 9. QUORUM. A majority of the whole number of Directors shall constitute a quorum at any meeting. In the absence of a quorum, a lesser number may adjourn any meeting from time to time, and the meeting may be held as adjourned, without further notice, if a quorum is obtained.

SECTION 10. EXPENSES. When the Directors meet for the transaction of

Association business their expenses incurred for such meetings may be paid from the funds of the Association, as the Directors decide at each meeting.

SECTION 11. ACTION WITHOUT A MEETING. Any action, which may be taken at a meeting of the Directors or of a committee, may be taken without a meeting if consent in writing setting forth the action so taken shall be signed by all of the Directors or all of the members of the committee entitled to vote thereon. Members of the Board of Directors may participate in a meeting of the Board or committee by means of conference telephone or similar communications equipment, by which all persons participating in the meeting can hear each other at the same time. Such participation shall constitute presence in person at the meeting.

ARTICLE IV
OFFICERS

SECTION 1. OFFICERS. The officers of the Association shall consist of the President, a Vice-President, Secretary/Treasurer, and such other officers as the Board of Directors deem necessary.

Officers shall be elected by the Board of Directors at the Board's Annual Meeting, and shall serve for a term of one year or until their successors are elected and qualified.

SECTION 2. PRESIDENT. The President shall be the Chief Executive Officer of the Association, and shall preside at all meetings of the Board of Directors and members; shall be ex-officio member of all committees; shall maintain general supervision of the affairs of the Association; shall see that the Bylaws and Rules and Regulations of the Association are enforced; shall have a vote in the Board of Directors in case of a tie; and, shall perform such other duties as may be prescribed by the Board of Directors.

SECTION 3. VICE-PRESIDENT. In the absence of the President, the Vice-President shall have the powers and shall perform the duties of the President, and shall perform such other duties as may be prescribed by the Board of Directors.

SECTION 4. SECRETARY/TREASURER. The Secretary/Treasurer shall keep or cause to be kept exact minutes of the meetings of the Board of Directors of the Association and shall perform such duties as directed by the President and by the Board of Directors. All secretary/treasurer records must be turned over to the registrar at the end of each year.

SECTION 5. REGISTRAR. A Registrar shall be employed by the Board of

Directors to receive and verify entries for insertion in the Herd Book subject to the Rules and Regulations of the Association; shall keep on file all documents constituting the authority for pedigrees and hold them subject to the inspection of any member of the Association; shall keep and be custodian of the funds and securities of the Association; and, shall deposit, invest or otherwise dispose of same as the Board of Directors may order; shall sign checks issued by the Association; and, shall perform all other duties properly ordered by the President or the Board of Directors, or which should be pertained to the office of the Registrar.

ARTICLE V
DISCIPLINE, SUSPENSION, EXPULSION

SECTION 1. VIOLATIONS. Whenever any members of the Association or any other person in interest shall represent to the Secretary of the Association in writing stating the facts upon which the complaint is based, that a member of this Association, or any other person who is a holder of a Certificate of Registration issued by this Association, has engaged in misrepresentation or misconduct in connection with the breeding, showing, registration, purchase or sale of _____, or has willfully violated the Bylaws, Rules and Regulations of this Association, the Secretary shall present such charge to the Board of Directors at its next meeting.

SECTION 2. HEARING. Upon receiving a complaint, the Board of Directors shall set a time and place for hearing the charge or charges against the member or holder of a Certificate of Registration. The Board of Directors shall cause a written notice to be mailed to the last known address of the accused person at least thirty (30) days before the date of such hearing. The notice shall state the nature of the charges against the accused.

At the time and place set for the hearing, the accused shall have the opportunity, in person or by counsel, to be heard and to present evidence in their own behalf and to hear and refute the evidence offered against him.

The decision of the Board of Directors shall be final and binding on all parties.

SECTION 3. PENALTIES. If the Board of Directors considers that the charges are sustained, it may suspend or expel such offender if a member of the Association, or impose such other appropriate penalties as it may decide and deprive him of all privileges in the official Record of the Association, including refusal to transfer any Certificate and Registration issued by this Association and cancellation of any registration of an animal standing in the name of the accused person. The Board, in its discretion, may also suspend and hold in abeyance during the pending of any complaint before it, the privileges of membership in the Association if the accused is a member of the Association, or the right to transfer any Certificate of Registration, if the accused is not a member.

ARTICLE VI
MISCELLANEOUS

SECTION 1. ORDER OF BUSINESS. The order of business of an Annual Meeting shall be:
- a) Calling the meeting to order by the President.
- b) Reading minutes of previous meeting and acting thereon.
- c) Annual address of the President.
- d) Reports of committees and old business.
- e) Election of directors.
- f) Unfinished business.
- g) New business.

In determining questions not covered by the Articles of Incorporation and Bylaws of this Association, Robert's Rules of Order shall be used. The order of business of the Directors' meeting shall be the same as Article VI, Section 1, except that those parts that are not applicable will be omitted.

SECTION 2. FISCAL YEAR. The fiscal year for the Association shall commence on January 1 and end on December 31.

SECTION 3. BONDS. The Registrar or any other employee entrusted with monies of the Association shall be bonded and/or covered by fidelity insurance. Such bonds and/or insurance shall be in an adequate amount as set by the Board of Directors and shall be an expense of the Association.

SECTION 4. AUDIT. It shall be the duty of the Board of Directors to cause to be audited all claims upon the Association and to verify the accounts of the Registrar before they are submitted to the members.

SECTION 5. NOMINATING COMMITTEE. The Board of Directors shall appoint a nominating committee of three members. The nominating committee will evaluate candidates according to guidelines established by the Board of Directors. Such committee shall consider all available candidates for the directorships and offices to be filled at the forth-coming meeting and shall submit a slate of candidates for election. Such submission shall be deemed a nomination of each person named. The committee may recommend one or more than one candidate for each vacancy to be filled.

At an annual meeting of members, nominations may be made by members from the floor.

SECTION 6. PROHIBITION AGAINST POLITICAL ACTIVITIES. The Corporation shall not participate or intervene in (including the publishing or distribution of statements) any political campaign on behalf of any candidate for public office.

SECTION 7. DISTRIBUTION OR DISSOLUTION. In the event of the dissolution of the Corporation, no member shall be entitled to any distribution or division of its remaining property or its proceeds, and the balance of all money and other property received by the Corporation from any source, after the payment of all debts and obligations of the Corporation, shall be used or distributed exclusively for the purposes within the intendment of Section 501c3 of the Internal Revenue Code as the same now exists or as it may be amended from time to time.

ARTICLE VII
AMENDING THE BYLAWS

These Bylaws may be altered or amended by a vote of the majority of the members of the Board of Directors in attendance at any Board meeting and confirmed by a majority vote of the membership voting.

These Bylaws may be amended by a two-thirds (2/3rds) vote of the qualified members voting in person at any annual meeting of the Association.

Proposed Articles of Incorporation or Bylaws changes must be presented in writing to the Board of Directors no less than two (2) months prior to the annual meeting. A proposed change in the Articles of Incorporation and Bylaws when approved by the Board of Directors will be published and forwarded to all members.

INDEX

Adalsteinsson, Stefan, 107
Adaptation
  definition/description, 107
  maintaining/selecting for, 107–110
African geese, 42, 85
Agricola Pineywoods cattle, 25, 107
Agricultural environments, 1–3
Akhal Teke horses, 103
ALBC. *See* American Livestock Breeds Conservancy (ALBC)
Alice in Wonderland, 128–129
Alpaca, 89–90
American Brahman cattle, 172–173
American Cream horses, 183–184
American Dairy Goat Association, 151
American Goat Society, 151
American Indian Horse Registry, 144–145
American Livestock Breeds Conservancy (ALBC)
  associations and, 150
  bloodlines and, 34–35
  card grading, 161
  education, 114, 136, 137, 138
  heritage turkeys, 114, 138
  long range conservation plans, 157
American Mammoth Jacks, 139, 180
American Nubian goats, 102
American Poultry Association (APA), 150, 165
American Quarter horses, 15, 133–134, 154
American Rabbit Breeders Association (ARBA), 150
American Sheep Industry, 150
Angora goats, 58
Angus cattle, 15, 23
Angus x Florida Cracker cattle, 115
Angus x Hereford cattle, 56–58
Appaloosa horses, 15
Arabian horses, 15
Arapawa goats, 126

Argentine Criollo horses, 91–92, 160
Artificial insemination
  for cattle, 17, 28, 89, 182
  inbreeding and, 17–18, 76, 78–79
  overview, 84
Associations
  board of directors, 148–149, 194
  breeder recruitment/training by, 138–142
  bylaws, 147–148, 192–199
  children/youth and, 130, 186
  Codes of Ethics, 135–136, 197–198
  communication function, 128–129, 131–133, 157
  directors/officers, 149, 194, 196–197
  education role, 128, 131, 136–138, 140, 142, 143, 152
  emergencies/rescues and, 132, 155–156
  forms of, 144–147
  importance/purpose summary, 127–128
  inbreeding and, 78–79
  membership, 129–130, 192–193
  multiple breed associations, 133–135
  networks of, 150
  politics/effects, 127, 128, 131–132, 133–135, 136, 145–146
  for rare breeds (summary), 130
  reputation of, 151
  research, 138
  responsibilities, 152–157
Auroch, 3

Bakewell, Robert, 39
Barbados Blackbelly sheep, 124–125
Barb horses, 21, 134
Belgian horses, 183
Biodiversity
  breeds and, 1, 2, 3, 4, 7, 111
  See also Genetic diversity

Index

Biology of breeds, 7–8
"Black baldy" cattle/matings, 56–58
Blanco Orejinegro cattle, 116
Bloodlines
    culling effects, 154
    "foundation" bloodlines, 15
    imported animals and, 120–121, 122–124
    linebreeding and, 67–69
    upgrading and, 105–106
    within breeds, 34–35
Bloodtyping, 22–23, 86, 168
Board of directors (associations), 148–149, 194
Boer goats, 124
Bottlenecks. *See* Genetic bottlenecks
Breeder responsibilities, 185–186
Breeding strategies
    goals and, 60–62
    for Java chickens, 31–32, 33
    over-representation of individuals, 80–83
    for population management, 65–66
    "related"/"unrelated" matings, 59–60
    See also specific strategies
Breeds
    definitions/description, 1, 7–8, 20
    evaluating, 20–23
    genetic organizations of, 27–34
    grouping/splitting of, 23–25
    See also specific breeds
Breed standards
    "breed type," 42, 47–50, 52
    changing, 51–52
    "defects" and, 44–46
    descriptions/development, 42–44
    descriptive standards, 39–40, 41–42
    genetic diversity loss and, 44–47
    of landraces, 39–40, 41–42
    prescriptive standards, 39, 40–41
    qualitative traits, 50–51
    quantitative traits, 50, 51
    of standardized breeds, 39, 40–41
Breed types
    breed standards and, 42, 47–50, 52
    modifications, 91–92, 96–97, 141, 159–160, 161
    selection and, 91–92
    See also "Type traits"/"typey"
Bylaws (associations), 147–148
    example, 192–199

Cabbage Hill Farm, New York, 118–119
Card grading, 161–163
Cashmere, 63
Caspian horses, 118
Cattle
    artificial insemination, 17, 28, 89, 182
    imports and, 122, 123
    mating systems, 87
    selection/effects, 88, 90
    wild ancestors of, 2–3
    See also Dairy cattle; *specific breeds*
Census, 179–182
Chickens
    egg production/genetic material management, 62–63
    industrial strains, 17, 28, 62–63
    See also Poultry breeds; *specific breeds*
Choctaw horses, 22, 25, 76, 87, 93, 145
Classes of breeds
    genetic diversity/uniformity summary, 19
    overview, 9–19
    summary, 9, 19
Cleveland Bay horses, 45, 62, 116, 153–154
Cloning, 83
Closed registry herd books, 174–175, 180
Clun Forest sheep, 172
Clutton-Brock, Juliet, 7
Clydesdale horses, 25, 120–121
Codes of Ethics (associations), 135–136, 197–198
Colonial Spanish horses
    associations for, 134–135, 144–145
    evaluating, 20, 21, 23, 40, 189–191
    feral populations, 19
    genetic bottlenecks, 83
    history, 36, 83
    imports and, 123–124
    mating systems, 87
    names for, 134
    registries, 176–177
    strains of, 22, 25, 36, 37, 145
    variability in, 39–40, 41
    See also specific breeds
Colonial Spanish horse score sheet (phenotype), 40, 189–191
Colonial Williamsburg Foundation, 68, 119
Columbia sheep, 14
Communication

associations and, 128–129, 131–133, 157
networks within associations, 131–133
Conflicts of interest (associations), 157–158
Congenital defects, 182–183, 184
Connemara ponies, 116
Conservation Breeding Handbook, A (Sponenberg and Christman), 65, 136, 137
Conservation of breeds
  breed evaluation and, 20–23
  genetic criteria of (overview), 8–9
  genetic organization of breeds and, 27, 31–34
  landraces and, 11–12, 13, 20
  linking populations with, 23–25
  long range conservation plans, 156–157
  See also Genetic diversity; *specific breeding strategies*
Conway Pineywoods cattle, 69, 76, 83, 88
Cotswold sheep
  Hillis Cotswold sheep, 69
  imports and, 104–105, 125–126
Creech, Cynthia, 81, 132
Criollo cattle, 10–11, 63, 64, 115–116
Criollo horses, 91–92, 160
Crossbreeding
  overview, 56–58
  purebreed decline with, 115–116
  threats from, 63–64
Cryopreservation, 156
Cryptorchidism, 157–158
Culling
  adaptations and, 108
  carriers of traits and, 45–46, 116–117, 153–154, 183–184
  inbreeding/harmful traits, 56
  off-standard traits and, 45–46, 153–154
  scrapie and, 46–47, 116–117
  type and, 94

Dairy and Beef Councils, 150
Dairy cattle
  EPD data, 95
  industrial strains, 17
  monitoring/reporting defects, 182
  selection/effects, 89
  See also specific breeds
Dale Ponies, 25, 26

Dalmatian dogs, 47
Demand. *See* Market demand
Devon cattle, 122, 123
Dexter cattle, 8
Directors/officers (associations), 149, 194, 196–197
DNA fingerprinting, 22–23
DNA testing
  evaluating breed, 22–23
  genetic defects and, 153, 154
  pedigrees and, 86, 168
Dominique chickens, 143, 166
Donkeys
  mating, 84–85
  See also specific breeds
Doroteo Baca New Mexico horses, 25, 37
Dorper hair sheep, 124–125
Drowns, Corey, 130
Duroc/Record Association, 135
Duroc swine, 135
Dutch Belted cattle, 117, 123

EBV (estimated breeding value), 94–95
Education
  associations and, 128, 131, 136–138, 140, 142, 143, 152
  breed promotion, 143
  training new breeders, 140, 142
Effective population size
  definition, 73
  determining, 73–74
  monitoring, 73–75
Eggerton, Sarah, 106–107
Embryo splitting, 84
Embryo transfer, 78–79, 84
Enderby Island cattle, 83
English Longhorn cattle, 101
EPD (estimated progeny difference), 94–96
Estimated breeding value (EBV), 94–95
Estimated progeny difference (EPD), 94–96
Evaluating breeds, 20–23
Exhibitions (noncompetitive) vs. shows, 163–164
Exmoor ponies, 45, 126

Fell ponies, 25, 26
Feral goats, Britain, 63
Feral populations, 18–19
Finnsheep (Finnish Landrace), 9, 172
"Flocks," 5

Florida Cracker cattle
    crossbreeding and, 67, 115, 143
    founders and, 10
    history, 36
    registries, 177
    strains, 63–64
    threats to, 115, 125
Florida Cracker horses, 25, 134
"Foundation" bloodlines, 15
Founder effect
    controlling, 82
    definition/description, 80–82
    examples, 81, 82, 119
    genetic bottlenecks and, 83
    landraces and, 9, 10–11, 30, 31
Friesian horses, 45, 153–154
Future of rare breeds, 1–2, 5–6, 186–188

"Gaggles," 5
Galloway cattle, 122
Galloway x White Shorthorn cattle, 61
Geese
    imports and, 118–119
    mating, 85
    See also specific breeds
Gene flow
    into/out of breeds, 96–97
    See also specific breeding strategies
Generation interval, 75–77
Genetic analysis of breed, 20, 21, 22–23
Genetic bottlenecks, 17, 36, 73, 76–77, 83, 84, 110, 123
    overview, 83
Genetic defects
    congenital defects vs., 182
    monitoring, 153–154, 182–184, 185–186
Genetic diversity
    in classes of breeds (summary), 19
    description, 2–3
    reporting, 155
Genetic diversity loss
    artificial insemination, 17, 28
    breed standards and, 44–47
    industrial strains, 17–18, 28, 90
    landraces, 11, 12–13
    modern "type" breeds, 16
    pedigrees and, 168–169
    production environments and, 4, 6
    standardized breeds, 14–16
    See also specific breeding strategies
Genetic isolation threat, 174–175, 186–187

Genetic management of breeds
    for long-term survival, 110
    See also specific strategies
Genetic organization of breeds
    conservation and, 27, 31–34
    description, 27–34
Geography and source herds, 37–38
Giant Chinchilla rabbits, 150
Goats
    associations for, 151
    dairy goats' upgrading/effects, 102
    imports and, 124, 126
    mating systems, 84–85, 87
    See also specific breeds
Guinea dwarf Pineywoods cattle, 88
Gulf Coast sheep, 12, 33–34
Gyr cattle, 48

Hamilton, Debbie, 119
Hampshire sheep, 96–97
Handspun Treasures from Rare Breed Wools, 115
Health issues
    associations and, 152–154, 184
    congenital defects, 182–183, 184
    genetic defects, 153–154, 182–184, 185–186
    monitoring/reporting, 153–154, 182–184, 185–186
    See also specific issues/problems
Herd books
    closed registry, 174–175, 180
    open registries, 175–178, 180
    terminology/description, 14
"Herds," 5
Hereford x Angus cattle, 56–58
Hickman Pineywoods cattle, 87, 88
Highland cattle, 122
Hillis Cotswold sheep, 69
Histories of breeds
    breed evaluation and, 20–21, 22, 23, 24–25
    importance of understanding, 35–36
Holderread's Waterfowl Conservation Center, Oregon, 119
Holstein cattle, 28, 74, 78, 89, 102, 116, 169
Holt Pineywoods cattle, 32
Horse of the Americas, 39–40, 145
Horses
    imports and, 118, 120–121, 123–124, 126

mating systems, 84–85, 87
stud reports, 171
Warmblood breeds, 8, 16, 116, 172
wild ancestors of, 2–3
See also specific breeds
Howard, Bo, 107
Hybrid vigor, 31, 56–57, 58, 61, 62, 70
HYPP ("Impressive Syndrome") trait, 154

Icelandic sheep, 125
Imported animals/breeds
    bloodlines and, 120–121, 122–124
    breed replacement by, 124–126
    conservation contribution by, 118–120
    negative effects/America, 122–126
    negative effects/country of origin, 121–122
    overview, 118
"Impressive Syndrome" (HYPP) trait, 154
Inbreeding
    across breeds, 71–72, 77–79
    controlling, 78–79
    effective population size and, 74
    individuals and, 71–72
    monitoring, 71–73
    overview, 53–56
    selection and, 88
    summary, 80
    upgrading and, 101–102
    within individual herds, 77
    See also Founder effect; Genetic bottlenecks
Inbreeding depression, 31, 66, 69–70
Incorporated associations, 146–147, 148
Industrial strains
    chickens, 17, 28, 62–63
    overview, 17–18, 19
    turkeys, 17, 90
International Species Identification System (ISIS), 170
Irish Draught horses, 116
ISIS (International Species Identification System), 170

Jacob sheep, 123, 186
Java chickens, 31–32, 33, 102
Jersey Buff turkeys, 86
Jersey cattle, 95
Jersey-Duroc/Record Association, 135
Junctional epidermolysis bullosa, 183–184

Karakul sheep, 123
Kensing line, Spanish goats, 103, 110
Kensing, Robert, 110
Kiko goats, 124
Kohlberg, Nancy, 118

Ladner Pineywoods cattle, 87
Landraces
    breed standards and, 39–40, 41–42
    definition/description, 9
    environmental effects, 9, 10, 12
    founder effect, 9, 10–11, 30, 31
    geography and, 37
    isolation effects, 9, 10, 11, 30
    modifications to, 12–13, 35
    overview, 9–14, 19
    registries, 174, 175, 176–177, 180
    See also specific breeds
Landrace swine breed, 9
Lang, Phil, 81
Leicester Longwool sheep
    breed standards, 39, 40
    environmental effects on, 2
    exhibition of, 163
    linebreeding and, 68–69
    promotion of, 143
    upgrading/imported animals, 104–105, 119–120, 121, 125–126
Lely strain, Texas Longhorn cattle, 55
Lincoln sheep, 14, 104–105, 125–126
Linebreeding
    applications of, 67–69
    outcrossing vs., 60–65
    overview, 53–55
Linebreeding/linecrossing alternation, 64–65, 79–80
Linecrossing, 56, 58
Linecrossing/linebreeding alternation, 64–65, 79–80
Litter recording, 170–171
Longwool sheep, 62

Malmberg, Peter, 31
Mammoth Jacks, 139, 180
Mangalarga horses, 124
Market demand
    breed survival and, 5, 111–115, 141
    products/services overview, 112
    rarity and, 112–113
    specialty markets, 113–115, 137, 138
Marsh Tacky horses, 25, 139
Mating systems/purebreds

overview, 84–87
pedigrees and, 84, 85, 86, 87
Merino sheep, 2, 29
Milking Shorthorn cattle, 8
Mission statements, 128
Mitochondrial DNA, 98
Modern "type" breeds, 16, 19
Monitoring
    breed health, 153–154, 182–184, 185–186
    effective population size, 73–75
    inbreeding, 71–73
    populations, 179–182
Morgan horses, 92, 141
Mount Taylor, New Mexico horse, 36, 37
Multiple breed associations, 133–135
Multi-sire mating systems, 86–87
Myotonic Goat Registry, 145–146, 177

National Animal Germplasm Program, 157
National Pedigreed Livestock Association, 150
National Turkey Federation, 150
Navajo-Churro sheep
    competitive showing, 160
    imports and, 125
    Navajo culture and, 112, 140
    registries, 176
    upgrading and, 103
    variation in, 45, 155
Newsletters, 131
Nokota horses, 23
Non-profit/not for profit status (associations), 146–147
Noriker horses, 117
Nubian goats, 49, 50, 102

Officers/directors (associations), 149, 194, 196–197
Open herd book registries, 175–178, 180
Ossabaw hogs, 18, 171
Outbreeding. *See* Outcrossing (outbreeding)
Outcrossing (outbreeding)
    linebreeding vs., 60–65
    overview, 56–58, 70
    summary, 80
Over-representation of individuals, 80–83

Paint horses, 15
Parasite resistance, 108–109, 116
Paso Fino horses, 124

Pedigrees
    association responsibilities, 152
    definition/description, 84, 166
    DNA testing, 86, 168
    genetic diversity loss, 168–169
    mating systems and, 84, 85, 86, 87
    monitoring inbreeding, 71, 84
    obtaining information on, 168
    recording/database systems for, 169–170
    registries vs., 167–168
    "related"/"unrelated" matings example, 59
Peruvian Paso horses, 124
Phenotype evaluation, 20, 21–22, 23, 40, 189–191
    Colonial Spanish horse score sheet, 40, 189–191
Pineywood oxen, 6, 187
Pineywoods cattle
    bloodlines and, 34–35
    crossbreeding, 67, 115, 143
    environmental effects on, 4
    founders and, 10
    genetic organization of, 30–31, 33
    grouping/splitting, 24–25
    modifications, 12, 13, 24–25
    registries, 177
    strain relationships, 31
    strains, 24–25, 31, 32, 63–64, 69, 76, 83, 87, 88, 107
    upgrading, 103
    variants, 155
Pinto horses, 7, 8, 16
Poitou donkeys, 100, 119, 120
Pomeranian geese, 52
Ponies
    registration of crossbreds, 178
    See also Horses; *specific breeds*
"Popular sire" phenomenon, 83
Populations
    monitoring/census, 179–182
    multiple breed associations and, 133, 134
    terminology, 5
    See also Effective population size
Pork Council, 150
Poultry breeds
    census of, 180–181
    flock validations, 165–166
    industrial strains, 17, 18
    standards, 40–41

See also Chickens; *specific breeds*; Turkeys
Predictability-variability balance, 25–27, 70
Private associations, 144–146
Promotion
    of associations, 151
    of breeds, 143–144
Pryor Mountain horses, 21, 25
Purebreds
    crossbreeding and, 62, 63
    description, 16
    mating systems/animal pairing, 84–87
    See also Upgrading

Quarter horses, 15, 133–134, 154

Rabbits
    associations for, 150
    registration, 170, 171, 182
Rambouillet sheep, 14
Randall cattle
    founders and, 72, 81, 82
    genetic bottlenecks, 83
    inbreeding, 55, 76, 81
    linebreeding, 54, 72
    rescue of, 132
Rare Breeds Survival Trust, UK, 161
Red Angus cattle, 23
Red Poll cattle, 5, 112, 113
Registries
    of crossbreds/partbreds, 178
    importance of, 172–174
    as inclusive, 130
    litter recording, 170–171
    monitoring inbreeding, 71
    monitoring populations, 179–182
    overview, 165–166
    pedigrees vs., 167–168
    selective recording systems, 171–172
    See also specific types
Regulations/effects, 117
Reproductive technologies
    overview/effects, 84
    See also specific technologies
Research needs/use, 138
Ridley Bronze turkeys, 69
Robert's Rules of Order, 149
Rocky Mountain horses, 43
Romeldale sheep, 126
Romo Sinuano cattle
    of Columbia, 10, 125
    of Venezuelen, 125
Royal Palm turkeys, 137

Sale events, 143–144
Salers cattle, 122
"Save the Sheep" program, 114–115
Saxony ducks, 181
Schell, Luther, 107
Scrapie
    overview, 46–47
    regulation and, 46–47, 116–117
Selection
    definition/importance, 87–88
    degree of/effects, 88–91
    genetic defects/carriers and, 153–154
    for improvement, 92–94
    type and, 91–92
    See also Culling
Sheep
    imports and, 119–120, 121, 123, 124–125, 125–126
    mating, 85
    specialty markets, 113, 114–115
    See also Scrapie; *specific breeds*
Shetland geese, 118–119
Shetland sheep, 126
Shire horses, 25, 120–121
Shorthorn cattle, 174
Shows
    card grading, 161–163
    effects summary, 159–160
    non-competitive exhibition vs., 163–164
    type changes and, 92, 97, 141, 159–160, 161
Single-sire group mating system, 85–86
Sitting Bull, Chief, 23
Slate turkeys, 156
Slow Food USA, 114
Soay sheep, 126
Society for the Preservation of Poultry Antiquities (SPPA), 150
Sorraia horses, 123–124
Southwest Spanish Mustang Association for the Colonial Spanish horse, 39–40
Spanish Barb horses, 134
Spanish goats
    breed purity and, 96
    imports and, 124
    Kensing line, 103, 110
    registries/pedigrees and, 167

Spanish Mustang Registry, 39–40
Spanish Mustangs, 19, 21, 134
Species associations, 150–151
Spinoff magazine, 114
Standardized breeds
    breed standards, 39, 40–41
    genetic diversity and, 14–16
    overview, 14–16, 19
St. Croix sheep, 124–125
Stud reports, 171
Suffolk sheep, 96–97
Sustainability of breeds (overview), 5–6
Swine
    industrial strains, 17, 18
    lard-type hogs, 46, 47
    litter recording, 171
    See also specific breeds

Tacky horse breed, 134
Tax status (associations), 146–147
Teeswater sheep, 104–105, 125–126
Tennessee Myotonic goats
    adaptations of, 108–109, 143
    association/registry for, 145–146, 177
    breeding strategies, 67, 70, 71
    geography and, 37
    imports and, 124
    upgrading, 106–107
Texas Longhorn cattle
    breed identity, 50
    crossbreeding, 115
    founders and, 10
    genetic bottlenecks, 83
    history, 36, 83
    inbreeding and, 55–56
    modification, 12, 35, 109, 110, 161
    registries, 174, 175–176
    strains, 55
Thoroughbred horses, 15, 102, 103, 116, 174
Tunis sheep, 123, 143

Turkeys
    inbreeding/heritage turkeys, 102
    industrial strains/effects, 17, 90
    marketing heritage turkeys, 114, 137, 138
    mating systems/heritage turkeys, 90
    selection and, 90
    See also Poultry breeds; *specific breeds*
"Type traits"/"typey"
    definition/description, 48, 49–50
    selection and, 92, 93, 94
    See also Breed types

Unincorporated associations, 146
Upgrading
    bloodlines and, 105–107
    breed consistency/predictability and, 99–100
    definition/description, 97–101
    inbreeding and, 101–102
    political aspects of, 101–102, 104–105
    recommendations, 103–104

Variability-predictability balance, 25–27, 70

Warmblood horse breeds, 8, 16, 116, 172
    See also specific breeds
Welsh Black cattle, 14–15
Welsh cattle, 14, 45
Welsh ponies, 23
Wensleydale sheep, 104–105, 125–126
Western Stock horse type, 15
White Holland turkeys, 130
White Park cattle, 44, 45
White Shorthorn x Galloway cattle, 61

About the Authors

D. Phillip Sponenberg (DVM from Texas A&M University, PhD from Cornell University) is Professor of Pathology and Genetics at the Virginia-Maryland Regional College of Veterinary Medicine. He teaches pathology, reproduction, genetic resources, and small ruminant medicine. Interests in coat color genetics include horses, donkeys, sheep, goats, dogs, and other species, and have resulted in publications in journals, book chapters, and books. He is active in rare breed conservation, and has had an active role as the technical advisor for the American Livestock Breeds Conservancy since 1977. His work has been pivotal in reversing the fate of several endangered breeds of American livestock. He and his wife Torsten maintain a herd of Tennessee Myotonic Goats in a wide variety of colors, and also own a Choctaw stallion.

Donald E. Bixby, (DVM from Michigan State University) served as ALBC's Executive Director from August 1988 through July 2002, Don was responsible for providing the vision and overseeing the implementation of the conservation activities of ALBC. Don has been involved with the organization since the 1980s, organizing the first North American rare breeds show and sale and establishing the ALBC Rare Breeds Gene Bank, which has expanded over the years. He has been the liaison to the USDA National Animal Germplasm Program and a leader in founding Rare Breeds International. He has overseen the livestock and poultry research, and promoted rare breeds to the sustainable agriculture community. He was honored in 2000 by Slow Food International for the work of ALBC in conserving genetic diversity in the farm animal species, and has been the ALBC liaison to the Renewing America's Food Traditions (RAFT) collaboration with six other food oriented non-profit organizations. He carries out his responsibilities as Technical Program Manager from Blacksburg, Virginia, where he lives with his wife Pat.